T0211295

SpringerBriefs in Applied Sciences and Technology

Series editor

Janusz Kacprzyk, Polish Academy of Sciences, Systems Research Institute, Warsaw, Poland

SpringerBriefs present concise summaries of cutting-edge research and practical applications across a wide spectrum of fields. Featuring compact volumes of 50–125 pages, the series covers a range of content from professional to academic.

Typical publications can be:

- A timely report of state-of-the art methods
- An introduction to or a manual for the application of mathematical or computer techniques
- A bridge between new research results, as published in journal articles
- A snapshot of a hot or emerging topic
- An in-depth case study
- A presentation of core concepts that students must understand in order to make independent contributions

SpringerBriefs are characterized by fast, global electronic dissemination, standard publishing contracts, standardized manuscript preparation and formatting guidelines, and expedited production schedules.

On the one hand, **SpringerBriefs in Applied Sciences and Technology** are devoted to the publication of fundamentals and applications within the different classical engineering disciplines as well as in interdisciplinary fields that recently emerged between these areas. On the other hand, as the boundary separating fundamental research and applied technology is more and more dissolving, this series is particularly open to trans-disciplinary topics between fundamental science and engineering.

Indexed by EI-Compendex and Springerlink.

More information about this series at http://www.springer.com/series/8884

Wouter Zijl · Florimond De Smedt
Mustafa El-Rawy · Okke Batelaan

The Double Constraint Inversion Methodology

Equations and Applications in Forward
and Inverse Modeling of Groundwater Flow

 Springer

Wouter Zijl
Hydrology and Hydraulic Engineering
Vrije Universiteit Brussel
Brussels
Belgium

Florimond De Smedt
Hydrology and Hydraulic Engineering
Vrije Universiteit Brussel
Brussels
Belgium

Mustafa El-Rawy
Department of Civil Engineering
Minia University
Minia
Egypt

Okke Batelaan
National Centre for Groundwater Research
    and Training
College of Science and Engineering,
    Flinders University
Adelaide, SA
Australia

ISSN 2191-530X   ISSN 2191-5318   (electronic)
SpringerBriefs in Applied Sciences and Technology
ISBN 978-3-319-71341-0   ISBN 978-3-319-71342-7   (eBook)
https://doi.org/10.1007/978-3-319-71342-7

Library of Congress Control Number: 2017962067

Printed on acid-free paper

This Springer imprint is published by Springer Nature
The registered company is Springer International Publishing AG
The registered company address is: Gewerbestrasse 11, 6330 Cham, Switzerland

Preface

Progress is the battle with the experts.

Peter Schlumbohm

Why another book with mathematical equations? I have already a book with these embellishments.

Anonymous geologist

This realization that the key to the understanding of Nature lay within an unassailable mathematics was perhaps the first major breakthrough in science.

Sir Roger Penrose

This work deals with the mathematical and applied aspects of the double constraint methodology to estimate the parameters that play a role in groundwater flow and is intended to serve students and practitioners by bridging the gap between basic hydrogeology and inverse groundwater modeling. Groundwater is world's largest freshwater source, but sound and sustainable exploitation remains a challenge. For the development and management of groundwater resources, a proper knowledge of the physical and mathematical laws and their parameters governing the state and movement of groundwater is essential. In the last decades, substantial progress has been made in inverse groundwater modeling, so that not only forward modeling, but also inverse modeling has become an essential tool for groundwater resources management. However, a good knowledge of basic mathematical principles of groundwater flow is essential to understand numerical models. Often such knowledge is lacking in traditional academic programs like civil engineering or geology. Hence, this work attempts to synthesize the mathematics of groundwater flow to give insight in the physics of the relevant parameters that characterize the porous formations through which groundwater flows. In this respect, this work provides in-depth information for geophysicists, hydrogeologists, and engineers pursuing Bachelor, Masters, and Ph.D. degrees, as well as for groundwater practitioners and consultants who intend to become skillful and competent modelers.

In addition, petroleum reservoir engineers and basin modelers will find ample inspiration to support their exploration and production-related modeling activities.

The inspiration for this work came from two sources. In the 2000s, the idea to apply the double constraint methodology to assist in the application of conventional data assimilation techniques was born in the oil and gas department of Netherlands Organisation for Applied Scientific Research TNO. Conventional data assimilation techniques are smoothing the subsurface permeability field in the flow models in such a way that geologists do no longer recognize their carefully constructed geological models. To mitigate the over-smoothing without destroying the models' match with measured pressures and flow rates, the double constraint methodology was one of the proposed methods. In the 1980s and 1990s, there was a unique experiment at TNO: managing director Prof. Frans Walter and his successor Dr. Hessel Speelman were encouraging petroleum engineers and scientists of the oil and gas department as wells as hydrogeologists and geoscientists of the ground-water department to join forces in research and development, preferably in coop-eration with universities. Subjects of common interest were modeling and uncertainty analysis. Among the many initiatives, research related to inverse modeling was initiated in close cooperation with the Department of Hydrology and Hydraulic Engineering of the Vrije Universiteit Brussel (VUB). Although TNO discontinued this experiment by end of the 1990s (from a commercial point of view, the petroleum market differs too much from the water market), the VUB team— Ph.D. students and supervisors—continued this research by developing mathe-matical proofs and models and performing case studies.

The reason why the double constraint methodology was initiated and continued at the VUB was that this methodology fits well as a research topic in its Water Resources Engineering Program for M.Sc. and Ph.D. students. This program, in which the Katholieke Universiteit Leuven (KU Leuven) has complementary tasks, is devoted to teaching and supervising water-related topics and is basically intended for international students, mostly from developing countries. The program has been running for a couple of decades and has close to a thousand graduates, working as researchers, consultants, academics, and practitioners all over the world. All stu-dents follow among others a course in groundwater hydrology, while electives as groundwater modeling have as ultimate goal to be proficient in modeling.

The students come from different backgrounds with diverse academic degrees in engineering, geological sciences, or environmental sciences. Engineers may have a profound knowledge of the mathematical–physical laws of conservation and movement, but often lack insight in specific properties and settings of groundwater-bearing formations and how these affect the equations describing groundwater flow. For earth scientists, it is usually the opposite as they have a good knowledge about geological conditions and processes, but often lack insight in the mathematical–physical aspects of describing flow and transport.

This above typical example of training students shows that better links are needed between basic geology/groundwater hydrology and groundwater modeling. The double constraint methodology, which is firmly based on the mathematical–physical laws of conservation and movement, as is amply explained and exemplified in this work, provides a link between "geologists" and "modelers." We therefore believe that this work contributes to a better understanding of groundwater flow theory, thus providing a greater and more realistic insight into what groundwater models can do and how they should be applied in practice.

Finally, we want to express our thanks to Dr. Anna Trykozko from the Interdisciplinary Centre for Mathematical and Computational Modelling, University of Warsaw, Poland. She played an important role in the initial phase of the development of the double constraint methodology, especially regarding stability and the question how to avoid negative conductivities. In addition, she has considerably improved this book, not only regarding text on stability and negative conductivities, but also regarding text on numerical over-estimation and the related difference between calibration and imaging.

Brussels, Belgium   Wouter Zijl
Brussels, Belgium   Florimond De Smedt
Minia, Egypt   Mustafa El-Rawy
Adelaide, Australia   Okke Batelaan

Contents

Chapter 1
Introduction: Setting the Scene

Groundwater flow models are necessary tools to understand a groundwater system, to make predictions about the system's response to a stress, or to design and support management interventions and decisions. In groundwater flow modeling, we generally distinguish two types of problems:

(i) The *forward* problem, in which the parameters (e.g., hydraulic conductivities) are specified, as well as the appropriate boundary and initial conditions.
(ii) The *inverse* problem, in which not all parameters are specified. Instead, additional boundary conditions, more than required for a forward problem, are imposed.

Inverse problems are common to many fields of sciences, such as geophysical and medical imaging, meteorological forecasting, petroleum reservoir engineering, and hydrology. Each time when we need to determine unknown properties of a system from the observations of a response of that system, inverse models come into play. Usually, the unknown properties are physical parameters characterizing the model and their values are determined by systematically adjusting them while checking the match between the model outputs and the observed parameters.

Groundwater flow models are based on the following two mathematical basic equations:

(i) The water balance equation: a partial differential equation describing the physical law of mass conservation—mass cannot be created or destroyed in the groundwater flow field.
(ii) The momentum balance equation: a partial differential equation describing the physical law of conservation of momentum. In most practical cases, the low Reynolds number of groundwater flow allows simplification of the law of conservation of momentum to Darcy's law: Flow rate is equal to hydraulic conductivity times head gradient.

© The Author(s) 2018
W. Zijl et al., *The Double Constraint Inversion Methodology*,
SpringerBriefs in Applied Sciences and Technology,
https://doi.org/10.1007/978-3-319-71342-7_1

For further explanations of the basic equations for groundwater flow see, for instance, Bear (1972, 1979), Bear and Verruijt (1987), Strack (1989), and for transport phenomena (not discussed in this book) coupled with groundwater flow see Bear and Corapcioglu (1985) and Bear and Cheng (2010).

In this book, the above-mentioned partial differential equations will be presented and applied in a more heuristic way, just to show the argument, for instance by replacing the space–time continuum by a discretized (or quantized) space–time to avoid dealing with functional analysis and infinite set theory. By doing so, we have tried to avoid an overly mathematical presentation. Regarding flow through porous media, the more rigorous mathematical treatment—function spaces, proofs of existence, uniqueness, consistency, stability, convergence, etc.—can be found in Chavent and Jaffré (1986). For more general applications, the more rigorous mathematics can be found in Butkov (1973) and Duvaut and Lions (1976). For the rigorous mathematics related to the inverse problem of the potential equation, to which we will devote attention in Chap. 2, Sects. 2.2 and 2.4, see Salo (2008) and the references in his paper.

Solutions to the water and momentum balance equations are characterized by the following two principles:

(i) If the hydraulic conductivity is specified over the full model domain, the two above-mentioned basic equations lead to a unique solution in the model domain only if they are complemented by initial and boundary conditions. More specifically, the full boundary enclosing the model domain may be partitioned in parts on which only one boundary condition may be imposed: either the head, or the flux (flow rate), or a linear combination of head and flux. This type of modeling is called forward modeling.

(ii) If we know more than one condition on one or more parts of the boundary, these additional boundary data can be used to determine the hydraulic conductivity in the model domain. Because of the lack of a sufficient number of measured data—heads and, especially, fluxes—the thus determined hydraulic conductivity field is generally not unique; there are generally a number of "equivalent hydraulic conductivity fields" that honor the imposed boundary conditions. This type of modeling is called inverse modeling.

We can distinguish two types of inverse modeling: (i) imaging and (ii) calibration. Imaging is the estimation of parameters, for instance hydraulic conductivities, as they occur in the physically existing real subsurface. Calibration, on the other hand, is the estimation of model parameters, e.g., hydraulic conductivities, such that the model matches with measured data like heads. For a more precise explanation of the difference between imaging and calibration see Sect. 3.9 of Chap. 3. Therefore, calibration is sometimes called parameter fitting and, in the context of processes evolving in time, data assimilation, or history matching (the latter term is generally used in petroleum reservoir simulation). Under the requirement that the model represents the physics of groundwater flow in a reliable way, the difference between imaging and calibration has to be relatively small. For such models, the difference may be considered as unimportant, especially for the practical case in which measurement errors determine the accuracy of the inversion. Therefore, the terms imaging and calibration will be used here as synonyms, except when the difference is stated explicitly. Also the more neutral terms parameter estimation and inversion have been used.

In the context of inversion, we will consider wells as internal boundaries: small holes in the porous subsurface on the boundary of which flow and/or head conditions are imposed as boundary conditions. An observation well is then considered as a well in which the head is known from measurements while the production/injection rate is equal to zero. To be more specific, this way of looking at wells is introduced for the explanation of the double constraint methodology (see below) and does not contradict the more practical modeling approach in which well flow rates are represented by source or sink terms in the field equations. The two approaches—internal boundaries and source/sink terms—are mathematically equivalent (Chavent and Jaffré 1986: 70–88). On the other hand, imposing wellheads cannot be accomplished by source/sink terms.

This book focuses on the estimation of hydraulic conductivities in discretized (or quantized) models, numerical models based on discretization of the continuum equations by means of a grid (mesh). In such models, a head has to be imposed in the center of the grid volume in which the well is situated. When dealing with an observation well (a piezometer), the imposed head is equal to the measured head (the piezometer's head). However, when dealing with a production/injection well, the grid volume head to be imposed has to be related to the measured head by a near-well model, generally a relatively simple algebraic model (Peaceman 1977, 1983, 1991; Chen and Zhang 2009).

Estimation of the parameters in a groundwater flow model is generally the most crucial step in the modeling process (Anderson and Woessner 1992; Anderson et al. 2015). Parameter estimation can be performed by a series of methods, ranging from manual calibration to complex automatic data assimilation algorithms (Frind and Pinder 1973; Neuman and Yakowitz 1979; Carrera and Neuman 1986a, b; Ginn and Cushman 1990; Certes and de Marsily 1991; Dykaar and Kitanidis 1992; Poeter and Hill 1997; Datta-Gupta et al. 1998; Gupta et al. 1998; Yeh and Liu 2000; Liu et al. 2002; Sun 2004; Farcas et al. 2004; Zhou et al. 2014). The problem of inverse modeling of groundwater flow has been studied extensively during the last four decades (de Marsily et al. 2000) and has been reviewed by Yeh (1981, 1986), Kuiper (1986), Carrera (1988), Keidser and Rosbjerg (1991), Sun (2004), Carrera et al. (2005), and Nilsson et al. (2007). Parameter estimation is the most challenging and time-consuming task of groundwater flow modeling, especially when the number of unknown parameters is large (Hill and Tiedeman 2007; Cao et al. 2006). It involves optimizing the model parameters to honor measured groundwater heads, fluxes, or concentration (McLaughlin and Townley 1996; Kitanidis 1997; Carrera et al. 2005; Pinault and Schomburgk 2006).

Zhou et al. (2014) present an overview of inverse modeling techniques in which they explain the traditional distinction between direct and indirect inverse modeling techniques. According to their definition, a numerically stable direct inversion method to determine hydraulic conductivities (e.g., in the grid volumes of a discretized groundwater flow model) requires specification of heads in each point of the model domain (i.e., in all the grid volume centers). In practice, heads are measured in only a limited number of points and, leaving the other heads unspecified, may lead the direct modeling approach to become unstable. Also, Sun (2004) shows that with only measured heads and fluxes as input data, direct inversion may become numerically unstable as small errors in head data may lead to large errors in hydraulic conductivity. On the other hand, indirect inversion methods are based on an algorithm (or just on trial and error) to iteratively adapt the hydraulic conductivities in the forward model until the calculated heads match with the measured heads (generally measured in observation wells) and such approaches are inherently stable. In the indirect inversion approach, the model is generally based on imposed recharge fluxes on the model's top boundary (the so-called flux model).

The double constraint method presented in this book shows that there is no need to draw a sharp demarcation line between the two types of inverse methods. Although the double constraint method (DCM) is a direct method because the DCM is based on heads imposed in all grid volumes centers, these heads are calculated by a forward model in which the heads measured in the observation wells are imposed; see Chap. 3 for more details.

As is well known, hydraulic conductivities can be estimated only if nonzero fluxes (at least one nonzero flux) are imposed on the boundary, while head is imposed on the other parts of the boundary (Haitjema 2006; Poeter and Hill 1997). Note that in this context the boundary includes internal boundaries like wells. In fact, the double constraint methodology (see below) may be considered as a rigorous application of these principles, because in this methodology the heads in observation wells (zero flux wells) are used as internal boundary conditions for a forward model, the so-called head model. The parameter estimation problem becomes underdetermined if many hydraulic conductivities have to be estimated and only a few measured head and flux pairs can be imposed. The resulting parameter non-uniqueness has to be mitigated by putting additional constraints.

Indirect methods are based on one forward model, generally a model in which as many fluxes as possible are imposed (a so-called flux model). This model fed with estimated hydraulic conductivities results in calculated heads that generally differ from the measured heads. These differences are then used to modify the hydraulic conductivities and so on until the head differences are minimized. As has already been mentioned above, a strong point of an indirect method is that it can handle any type of parameterized process, independent from the mathematical equations and boundary conditions that govern the process (Sun 2004; Hill and Tiedeman 2007; Doherty and Hunt 2009, 2010; Doherty et al. 2010). Indirect inverse methods are versatile as any type of data as input, including soft data, can be used (Tikhonov 1963a, b; Doherty 2003; Hunt et al. 2007). Automatic parameter estimation is most often done with a gradient method. It has been used in, e.g., the pilot point and sequential self-calibration methods (Fasanino et al. 1986; RamaRao et al. 1995; LaVenue et al. 1995; Wen et al. 1998, 1999; Doherty 2003). As gradient methods have good convergence rates, they can find very accurately the parameters that match the measured data (Sun 2004; Hager and Zhang 2006; Ewald 2006).

In this book, we consider only heads and fluxes as measured data. However, when considering transport problems in which concentrations of dissolved matter are available, these measurements can yield useful additional information about the heterogeneity of the hydraulic conductivity field. In such cases, reduction of the heterogeneity by a predefined parameterization, such as zonation (see Chap. 6), is questionable, because such a parameterization may be too coarse to account for the fine-scale heterogeneity by which the transport process is strongly influenced. Seven geostatistical inversion methods were tested by Zimmerman et al. (1998) for how appropriate they are for making probabilistic forecasts of solute transport with heterogeneous hydraulic conductivity. Valstar (2001) presents a representer approach using measured heads complemented by measured concentrations to determine the fine-scale heterogeneity within a zone of the subsurface; also see Valstar et al. (2004). Generally, a representer approach is based on a series expansion in which the number of state variables and/or parameters is reduced to the number of measurements used by the inversion (Bennett 1992). In that sense, Eq. 2.10 in Chap. 2 may be considered as a relatively simple example of a representer approach.

Hendricks Franssen et al. (2009) review and compare seven recent inverse approaches; they show that the performance of the methods depends strongly on the type of heterogeneity. This is partly due to subjective decisions in translating the conceptual to the numerical parameter estimation processes (Rajanayaka and Kulasiri 2001). Another issue is that there appears to be no consistency on how to select the most appropriate method for a given problem.

Direct inverse methods, in contrast to indirect inverse methods, estimate hydraulic conductivities directly from Darcy's law, mostly while combining the water balance equation. Direct methods for estimating hydraulic conductivity fields appeared around the 1950s. One of the first practical applications was published by Huisman (1950) followed by Nelson (1960, 1961, 1962, 1968), Emsellem and de Marsily (1971), Neuman (1973) and Sagar et al. (1975). An equation stating the relationship between the log-conductivity field and the head field was derived by combining Darcy's law and the mass conservation equation, while neglecting specific and phreatic storage. It was concluded that only if the head is fully specified in the model domain, and if the flux density is given at boundaries, integration along the streamlines allows successful determination of the hydraulic conductivity field.

At present, gradient-based minimization of an objective function, i.e., the difference between observed and simulated data of a forward model, is probably the most popular and widespread approach to automatic parameter estimation (Sun 2004; Tarantola 2005). In principle, such gradient methods can be used to determine all types of parameters that occur in any mathematical model. Moreover, from a mathematical point of view gradient methods have the best convergence properties, at least if the initial estimate of the parameters is close to the parameters for which the forward model matches the measurements. However, gradient-based minimization has a number of practical limitations. The numerical complexity of handling the gradient matrix increases considerably when the number of spatial measurement points increases. This may lead to unacceptable high computational requirements (computer time and memory), especially if uncertainty estimation, which is based on multiple parameter estimation runs, is required. The adjoint method is the most efficient way to compute the gradient of a defined objective function (Li et al. 2003). However, the adjoint system requires modification of the source code of the computer model, which is time-consuming and often impossible when a commercial model is used. Furthermore, if we choose a different model the source code for the adjoint calculations must be recoded again.

The heavy computational burden, especially in problems dealing with big data, and/or the required modification of the source code for inverse modeling by a gradient method has motivated petroleum reservoir engineers and hydrologists to apply the ensemble Kalman filter (EnKF) (Evensen 1994; Naevdal et al. 2002; Chen and Zhang 2006; Aanonsen et al. 2009). The EnKF is a Bayesian approach to model updating and parameter estimation. The method uses Monte Carlo statistics for merging observation data with forecasts from model simulations to estimate a range of plausible models. The ensemble of updated models is then used to estimate the forecast and model parameters, as well as their uncertainty. Thanks to its simple formulation, its ability to account for measurement errors and model noise, and its relative ease of implementation for any simulator model, the EnKF has gained popularity. It requires neither the time-consuming computational handling of a gradient matrix nor the construction of adjoint equations. It supports efficient uncertainty assessment and integration of diverse data types. Because the EnKF still requires many runs—realizations from an ensemble—to provide stable estimates, its computational effort may be reduced by terminating the EnKF calculations after having reached a steady Kalman filter (SKF). The SKF is then used for the subsequent forecasts and parameter estimations. The EnKF-SKF approach is appreciably less demanding in computational effort while maintaining stable and reliable estimates (El Serafy and Mynett 2008).

Although parameters like hydraulic conductivity can be measured directly on the laboratory scale, they cannot be measured directly on field scales. On such scales, they have to be determined by groundwater flow models fed with directly and indirectly measured quantities. Directly measured quantities are, for instance, flow rates into or out of wells, pressures, or heads measured in wells and observation wells. Indirectly measured quantities, for instance recharge rates, are determined indirectly with the aid of models that interpret directly measured data, for instance meteorological data. In the sequel, indirectly measured data (mainly recharge rates) will be denoted as "measured" data (i.e., "measured" recharge rates).

Parameters are time-independent by definition. However, they may vary from point to point in continuous space, or from grid volume to grid volume in discretized (quantized) space. Because the number of head measurements and flux "measurements" is limited, it is impossible to find the spatially distributed parameter field that exists in reality. Instead of finding the only one real parameter field, we will generally find equivalent parameter fields, each of which will allow to match the calculated heads and fluxes to the measured heads and fluxes.

This book deals mainly with the double constraint methodology (DCM) for inverse modeling. The motivation to develop the DCM was the same as the motivation for the EnKF-SKF approach: Especially when dealing with big data models, we have to avoid the adjoint method (a gradient method) because of its heavy computational burden and/or its required modification of the source code. Although this book introduces the DCM mainly as a "stand-alone methodology," we present arguments why and how the DCM could be applied in combination with the EnKF-SKF (see Chap. 4, Sect. 4.3.4).

The double constraint methodology (DCM) is based on two forward ground-water flow models with the same initial hydraulic conductivity: (i) a model in which known (measured) fluxes are imposed in wells and on the boundary, and (ii) a model in which known (measured) heads are imposed in wells—including the observation wells—and on the boundary. The difference between head and flux obtained by the two models is then used to update the initial hydraulic conductivity. In this approach, Darcy's law is applied directly using heads that are estimated by a "head model" based on the relatively limited measured head data. Therefore, in view of the definition of direct methods presented by Zhou et al. (2014), it is better to characterize this method as an "indirect direct inversion method." A similar method was also tested as an approach to upscale hydraulic conductivities (Warren and Price 1961; Durlofsky 1991; Trykozko et al. 2001; Zijl and Trykozko 2001).

Like conventional gradient methods, DCM is based on minimization. To avoid the occurrence of negative hydraulic conductivities, DCM is formulated in terms of the square root of the hydraulic conductivity (in short the sqrt-conductivity). DCM's computational complexity is relatively modest compared to conventional approaches, even if each grid volume in a numerical model is considered as a zone with its own hydraulic conductivity (El-Rawy et al. 2015).

The inspiration to base the parameter estimation on a second model in which known (measured) heads are imposed came from groundwater flow systems analysis, in which the head or water table height is prescribed on the model's top boundary, rather than the recharge fluxes (Tóth 2009). Compared to the great step of this flow systems approach to modeling, it was a relatively small step to impose also the heads in the observation wells, rather than the zero fluxes, which are "imposed" automatically in the conventional flux model.

The double constraint methodology was first published in the 1980s in the context of electrical impedance tomography (EIT) for geophysical and medical imaging (Wexler et al. 1985; Kohn and Vogelius 1987; Wexler 1988; Kohn and McKenney 1990); for more background theory regarding EIT see Borcea (2002, 2003), Borcea et al. (2003) and Calderón (1980). The name double constraint method was coined by Yorkey and Webster (1987) and Yorkey et al. (1987), Webster (1990). These applications are based on finite element methods for electrical potentials and currents through n-dimensional ($n = 2, 3$) domains bounded by curved boundaries (e.g., the skin of a part of the human body). Although the method has been applied in the field of hydrogeology to simulate groundwater contamination (Tamburi et al. 1988), the methodology remained unnoticed in the geoscientific community until it was reinvented in the context of petroleum reservoir engineering (Brouwer et al. 2008) and, later on, in the context of hydrogeology (Trykozko et al. 2008, 2009; El-Rawy et al. 2010, 2011; El-Rawy 2013; El-Rawy et al. 2015, 2016). The subsequent geoscientific applications involved artificial numerical or experimental n-dimensional flow conditions (mostly with $n = 2$) flow in rectangular domains. Wexler et al. (1985) and Wexler (1988) found DCM to be promising; the method does not diverge, but convergence was sensitive to noise. Kohn and Vogelius (1987) proposed some numerical improvements, the most important one a DCM formulation that avoids negative (unrealistic) conductivity values. Tests by Kohn and McKenney (1990) still showed convergence instability (oscillations) when there is noise in the data. Kohn and McKenney (1990) concluded that it is not desirable to continue DCM iterations up to stable convergence; earlier termination has a smoothing effect on instabilities and is generally sufficiently accurate. Also, the DCM application described by Yorkey and Webster (1987) and Yorkey et al. (1987) showed similar convergence behavior. Brouwer et al. (2008) applied DCM in petroleum reservoir engineering to estimate permeability fields from pressure and flow rate measurements in wells. These tests involved the reconstruction of synthetic two-dimensional spatially correlated permeability fields and showed that DCM performs well in a few iterations near the wells while uncertainty remains in the regions further away from the wells. Brouwer et al. (2008) compared the thus-obtained permeabilities with the permeabilities obtained by the ensemble Kalman filter (EnKF). This comparison was also the starting point for application of the Kalman filter in combination with the DCM to obtain estimation uncertainties (see Chap. 4, Sects. 4.2 and 4.3 and Chap. 5). Trykozko et al. (2008, 2009) presented DCM applications for upscaling and downscaling of synthetic checkerboard hydraulic conductivity patterns and noted several anomalies as occurrence of negative hydraulic conductivity values or extremely high hydraulic conductivity values that tend to diverge during the iterations, as well as smoothing of large hydraulic conductivity contrasts, for which they recommended renormalization. El-Rawy (2013) and El-Rawy et al. (2015) presented a real-world DCM application (Kleine Nete basin, Belgium) by updating an initially specified hydraulic conductivity field and found that sufficiently accurate improvements are obtained near the measurement locations after a few iterations, usually 2–5 DCM runs, but no improvement is possible far away from the measurement locations. They concluded that DCM is much cheaper in computation time and memory requirements than indirect calibration methods like UCODE or PEST and, therefore, a valuable tool for calibration of groundwater models. El-Rawy (2013) and El-Rawy et al. (2016) presented another real-world inverse problem (Schietveld area, Belgium) with a DCM application very similar as in the previous case. The results were also very similar; i.e., DCM updates hydraulic conductivities only near the piezometers, while locations far away from the piezometers are hardly influenced. They also noted that DCM's better calibration performance is likely due to the number of degrees of freedom compared to the very limited heterogeneity and anisotropy patterns allowed by an indirect inversion procedure as UCODE.

In the double constraint method as presented by El-Rawy (2013) and El-Rawy et al. (2010, 2011, 2015, 2016), the hydraulic conductivity estimation was "pointwise" or, more precisely, "grid block wise." In Chaps. 6 and 7, we present an interesting extension, namely a zone integrated DCM. The goal of this newer approach is to enable inverse modeling with limited data, so that hydraulic conductivity estimates can be improved in the whole model domain instead of only near the observations. In order to do so, the model domain is partitioned in zones with presumed constant hydraulic conductivity (soft data) and DCM is reformulated accordingly for calibrating a steady-state groundwater model given a series of measured heads.

The DCM can be considered as a relatively simple minimization methodology that is not based on a gradient matrix. In DCM, the objective function to be minimized is a so-called Darcy residual based on a form of Darcy's law in which the square root of the hydraulic conductivity (the sqrt-conductivity) is used as parameter, rather than the hydraulic conductivity. As a consequence, the DCM cannot be used to determine all types of parameters. It can only be applied to the parameters in constitutive laws such as Darcy's law, Ohm's law, Fick's law, Fourier's law, Hooke's law. In addition, these laws have to be embedded in process equations that lead to a well-posed forward problem for more types of boundary conditions, as is the case for Darcy's law embedded in the water balance equation, which can be solved either with flux boundary conditions, or with head boundary conditions (for more details see Chap. 2). In the context of hydrogeology also specific yield, specific storage and dispersion coefficient can be determined by the DCM, at least in principle. On the other hand, gradient methods, as well as the ensemble Kalman filter (EnKF) and the steady Kalman filter (SKF), can do the job of determining any type of parameter.

Although the DCM has not the best convergence properties, it has the advantage that it improves the initial estimate of the parameters even if this estimate is far away from the parameters for which the forward model matches the measurements. Even more importantly, the numerical complexity of handling the minimization iterations is hardly increasing when the number of spatial measurement points and parameters to be determined increases. This may lead to acceptable computer requirements regarding computation time and memory requirements, especially if multiple inverse modeling runs, e.g., for uncertainty estimation, are required. In contrast to gradient methods and the EnKF/SKF, the DCM is less prone to over-smoothing the spatial parameter distribution.

To be more specific, the double constraint methodology is based on two models. One model—the "flux model"—uses as many boundary fluxes as possible, like the model that is generally used in indirect inversion methods. The other model—the "head model"—uses as many boundary heads as possible, which means that also the heads measured in the observation wells have to be imposed. The remainder of this book is devoted to the presentation and explanation of the double constraint methodology, as well as the Kalman Filter, which will be used to investigate the inversion accuracy by sequentially updating the DCM-determined hydraulic conductivities (El-Rawy 2013; El-Rawy et al. 2015).

This first chapter presents an introduction to the subject matter and some relevant literature. The second chapter explains how the physical laws of conservation of mass and conservation of momentum are applied to groundwater flow and enable to formulate the basic laws of groundwater flow. Special attention is paid to the parameters that occur in the equations, especially to the hydraulic conductivity and the important role of its square root. Over-smoothing and under-smoothing of the hydraulic conductivity field by inversion methods is addressed by considering the Calderón problem. The third chapter discusses the double constraint methodology and its different appearances in more detail, with a justification of the method. The fourth chapter focuses on time-dependent flow. Not only the parameters that play a role in the time-dependent terms of the equations, but also the decrease in uncertainty when time series of measurements are taken into account are discussed. The latter point, uncertainty versus observation error, is considered from a Bayesian point of view by application of the Kalman filter and is exemplified in Chap. 5 for the case study of the Kleine Nete basin. Chapter 6 presents an extension to the double constraint method to account for zonation, a popular parameter reduction technique, which makes it possible to compare the double constraint results with results obtained from conventional inversion techniques. Also, a brief account of the comparison with a conventional method for a realistic case is presented. Chapter 7 summarizes the results and presents some directions for further research and development. Chapters 2, 3, and 4 have a last section with the title beginning with "In depth: a closer look at …". These last sections focus more in depth on the theoretical aspects presented in the previous sections. Reading of these sections is not necessary for understanding the message of this book, but may be helpful for the curious reader looking for more background information.

The references at the end of the chapters are not intended to be comprehensive, but direct the reader to works that can be considered as supporting aids in the theoretical and practical development of the double constraint methodology.

References

Aanonsen SI, Naevdal G, Oliver DS, Reynolds AC, Vallès B (2009) The ensemble Kalman filter in reservoir engineering: a review. SPE J 14 (03). https://doi.org/10.2118/117274-PA

Anderson MP, Woessner WW (1992) Applied groundwater modeling: simulation of flow and advective transport. Academic Press, San Diego

Anderson MP, Woessner WW, Hunt RJ (2015) Applied groundwater modeling: simulation of flow and advective transport, 2nd edn. Academic Press, San Diego

Bear J (1972) Dynamics of fluids in porous materials. American Elsevier Publ. Co., New York

Bear J (1979) Hydraulics of groundwater. McGraw-Hill Book Co., New York

Bear J, Corapcioglu MY (1985) Fundamentals of transport phenomena in porous media. Kluwer Acad. Publ, Hingham, MA

Bear J, Verruijt A (1987) Modeling groundwater flow and pollution. Reidel Publishing Company, Dordrecht

Bear J, Cheng AH-D (2010) Modeling groundwater flow and contaminant transport. Springer, Netherlands

Bennett AF (1992) Inverse methods in physical oceanography. Cambridge Univ. Press, New York

Borcea L (2002) Electrical impedance tomography. Inverse Prob 18:99–136

Borcea L (2003) Addendum to electrical impedance tomography. Inverse Prob 19:9978

Borcea L, Gray GA, Zhang Y (2003) Variationally constrained numerical solution of electrical impedance tomography. Inverse Prob 19:1159–1184, PII: S0266-5611(03)57791-7

Brouwer GK, Fokker PA, Wilschut F, Zijl W (2008) A direct inverse model to determine permeability fields from pressure and flow rate measurements. Math Geosc 40(8):907–920

Butkov E (1973) Mathematical physics. Addison Wesley Publ. Comp., Reading

Calderón AP (1980) On an inverse boundary value problem, Seminar on Numerical Analysis and its Applications to Continuum Physics (Rio de Janeiro, 1980), Soc. Brasil. Mat. 65–73, Rio de Janeiro. Also see the reprint: Calderón AP (2006) On an inverse boundary value problem, Comput Appl Math 25(2–3):133–138

Cao W, Bowden WB, Davie T, Fenemor A (2006) Multi-variable and multi-site calibration and validation of SWAT in a large mountainous catchment with high spatial variability. Hydrol Process 20:1057–1073

Carrera J (1988) State of the art of the inverse problem applied to the flow and solute transport equations. In: Custodio E, Gurgui A, Ferreira JL (ed) Groundwater flow and quality modelling Publ. Comp., Netherlands: 549–583

Carrera J, Alcolea A, Medina A, Hidalgo J, Slooten LJ (2005) Inverse problems in hydrogeology. Hydrogeol J 13:206–222

Carrera J, Neuman SP (1986a) Estimation of aquifer parameters under transient and steady state conditions 1: Maximum likelihood method incorporating prior information. Water Resour Res 22(2):199–210

Carrera J, Neuman SP (1986b) Estimation of aquifer parameters under transient and steady state conditions 2: Uniqueness, stability and solution algorithms. Water Resour Res 22(2):211–227

Certes C, de Marsily G (1991) Application of the pilot points method to the identification of aquifer transmissivities. Adv Water Resour 14(5):285–300

Chavent G, Jaffré J (1986) Mathematical models and finite elements for reservoir simulation. Elsevier, North-Holland, Amsterdam

Chen Y, Zhang D (2006) Data assimilation for transient flow in geologic formations via Ensemble Kalman Filter. Adv Water Resour 29:1107–1122

Chen Z, Zhang Y (2009) Well flow models for various numerical methods. Int J of Numer Anal Model 6(3):375–388

Datta-Gupta A., Yoon S, Barman I, Vasco DW (1998, December) Streamline-based production data integration into high resolution reservoir models. J Pet Technol, 72–76

de Marsily G, Delhomme JP, Coudrain-Ribstein A, Lavenue AM (2000) Four decades of inverse problems in hydrogeology. Geological Society of America, Special Paper, p 348

Doherty J (2003) Ground water model calibration using pilot points and regularization. Ground Water 41(2):170–177

Doherty JE, Hunt RJ (2009) Two statistics for evaluating parameter identifiability and error reduction. J Hydrol 366:119–127. https://doi.org/10.1016/j.jhydrol.2008.12.018

Doherty JE, Hunt RJ (2010) Response to comment on: two statistics for evaluating parameter identifiability and error reduction. J Hydrol 380:489–496. https://doi.org/10.1016/j.jhydrol.2009.10.012

Doherty JE, Hunt RJ, Tonkin MJ (2010) Approaches to highly parameterized inversion: a guide to using PEST for model-parameter and predictive-uncertainty analysis. U.S. Geological Survey Scientific Investigations Report 2010–5211

Durlofsky LJ (1991) Numerical calculation of equivalent grid block permeability tensors for heterogeneous porous media. Water Resour Res 27:699–708

Duvaut G, Lions JL (1976) Inequalities in mechanics and physics, Springer-Verlag, Berlin, ISBN: 978-3-642-66167-9 (Print) 978-3-642-66165-5 (Online)

Dykaar BB, Kitanidis PK (1992) Determination of the effective hydraulic conductivity for heterogeneous porous media using a numerical spectral approach: 1 Method. Water Resour Res 28(4):1155–1166

Evensen G (1994) Sequential data assimilation with a nonlinear quasi-geostrophic model using Monte Carlo methods to forecast error statistics. J Geophys Res 99(C5):10143–10162

El-Rawy M (2013) Calibration of hydraulic conductivities in groundwater flow models using the double constraint method and the kalman filter. Ph.D. thesis Vrije Universiteit Brussel, Brussels, Belgium

El-Rawy M, Batelaan O, Zijl W (2015) Simple hydraulic conductivity estimation by the Kalman filtered double constraint method. Groundwater 53(3):401–413. https://doi.org/10.1111/gwat.12217

El-Rawy M, De Smedt F, Batelaan O, Schneidewind U, Huysmans M, Zijl W (2016) Hydrodynamics of porous formations: Simple indices for calibration and identification of spatio-temporal scales. Mar Petrol Geol 78:690–700. https://doi.org/10.1016/j.marpetgeo.2016.08.018

El-Rawy M, Mohammed GA, Zijl W, Batelaan O, De Smedt F (2011) Inverse modeling combined with Kalman filtering applied to a groundwater catchment. In: Proceedings of the MODFLOW and more conference, 6–8 June 2011, Golden, USA

El-Rawy M, Zijl W, Batelaan O, Mohammed GA (2010) Application of the double constraint method combined with MODFLOW. In Proceedings of the Valencia IAHR international groundwater symposium, 22–24 September 2010, Valencia, Spain

El Serafy GY, Mynett AE (2008) Improving the operational forecasting system of the stratified flow in Osaka Bay using an ensemble Kalman filter–based steady state Kalman filter. Water Resour Res 44(6):W06416. https://doi.org/10.1029/2006WR005412

Emsellem Y, de Marsily G (1971) An automatic solution for the inverse problem. Water Resour Res 7(5):1264–1283

Ewald CO (2006) The Malliavin gradient method for the calibration of stochastic dynamical models. Applied Math Comput 175(2):1332–1352

Farcas A, Elliott L, Ingham DB, Lesnic D (2004) An inverse dual reciprocity method for hydraulic conductivity identification in steady groundwater flow. Adv Water Res 27:223–235

Fasanino G, Molinard JE, de Marsily G, Pelce V (1986) Inverse modeling in gas reservoirs. SPE 15592, 61st SPE Annual Technical Conference and Exhibition, October 5–8, New Orleans, Louisiana

Frind EO, Pinder GF (1973) Galerkin solution of the inverse problem for aquifer transmissivity. Water Resour Res 9(5):1397–1410

Ginn TR, Cushman JH (1990) Inverse methods for subsurface flow: a critical review of stochastic techniques. Stochastic Hydrol Hydraul 4:1–26

Gupta VK, Sorooshian S, Yapo PO (1998) Towards improved calibration of hydrologic models: multiple and noncommensurable measures of information. Water Resour Res 34:751–763

Hager WW, Zhang H (2006) A survey of nonlinear conjugate gradient methods. Pacific J Optim 2 (1):35–58

Haitjema H (2006) The role of hand calculations in ground water flow modeling. Ground Water 44 (6):786–791. https://doi.org/10.1111/j.1745-6584.2006.00189.x

Hendricks Franssen HJ, Alcolea A, Riva M, Bakr M, Van der Wiel N, Stauffer F, Guadagnini A (2009) A comparison of seven methods for the inverse modelling of groundwater flow. Application to the characterisation of well catchments. Adv Water Resour 32(6):851–872

Hill MC, Tiedeman CR (2007) Effective groundwater model calibration: with analysis of data, sensitivities, predictions, and uncertainty. Wiley, New York

Huisman L (1950) Resistance of clay-layer Amsterdam dune water catchment area (in Dutch). Amsterdam Water Supply Report

Hunt RJ, Doherty JE, Tonkin MJ (2007) Are models too simple? Arguments for increased parameterization. Ground Water 45(3):254–261. https://doi.org/10.1111/j.1745-6584.2007.00316.x

Keidser A, Rosbjerg D (1991) A comparison of four inverse approaches to groundwater flow and transport parameter identification. Water Resour Res 27(9):2219–2232

Kitanidis PK (1997) The minimum structure solution to the inverse problem. Water Resour Res 33 (10):2263–2272

Kohn RV, McKenney A (1990) Numerical implementation of a variational method for electrical impedance tomography. Inverse Probl 6:389–414

Kohn RV, Vogelius M (1987) Relaxation of a variational method for impedance computed tomography. Comm Pure and Appl Math 40:745–777

Kuiper LK (1986) A comparison of several methods for the solution of the inverse problem in two-dimensional steady-state groundwater flow modeling. Water Resour Res 22(5):705–714

LaVenue AM, RamaRao BS, de Marsily G, Marietta MG (1995) Pilot point methodology for automated calibration of an ensemble of conditionally simulated transmissivity fields 2: application. Water Resour Res 31(3):495–516

Li R, Reynolds AC, Oliver DS (2003) Sensitivity coefficients for three-phase flow history matching. J Can Pet Technol 42(4):70–77. https://doi.org/10.2118/03-04-04

Liu X, Illman WA, Craig AJ, Zhu J, Yeh TCJ (2002) Laboratory sandbox validation of transient hydraulic tomography. Water Resour Res 38(4):1034–1043

McLaughlin D, Townley LR (1996) A reassessment of the groundwater inverse problem. Water Resour Res 32(5):1431–1161

Naevdal G, Mannseth T, Vefring EH (2002) Near-well reservoir monitoring through Ensemble Kalman Filter. Paper SPE 75235 (9 pages), SPE/DOE Improved Oil Recovery Symposium, 13–17 April 2002, Tulsa, Oklahoma, USA

Nelson RW (1960) In-place measurement of permeability in heterogeneous porous media 1: theory of a proposed method. J Geophys Res 65(6):1753–1758

Nelson RW (1961) In-place measurement of permeability in heterogeneous porous media 2: experimantal and computational considerations. J Geophys Res 66(8):2469–2478

Nelson RW (1962) Conditions for determining areal permeability distribution by calculation. Soc Petrol Eng J 2(3):223–224. http://www.onepetro.org/mslib/servlet/onepetropreview?id=00000371

Nelson RW (1968) In-place determination of permeability distribution for heterogeneous porous media through analysis of energy dissipation. Soc Petrol Eng J 8(1):33–42. http://www.onepetro.org/mslib/servlet/onepetropreview?id=00001554

Neuman SP, Yakowitz S (1979) A statistical approach to the inverse problem of aquifer hydrology 1: theory. Water Resour Res 15(4):845–860

Nilsson B, Højberg AL, Refsgaard JC, Troldborg L (2007) Uncertainty in geological and hydrogeological data. Hydrol Earth Syst Sci 11:1551–1561

Neuman SP (1973) Calibration of distributed parameter groundwater flow models viewed as a multiple-objective decision process under uncertainty, Water Resour Res 9(4):1006–1021

Peaceman DW (1977) Interpretation of well-block pressures in numerical reservoir simulation. SPE 6893, 52nd Annual Fall Technical Conference and Exhibition, Denver

Peaceman DW (1983) Interpretation of well-block pressures in numerical reservoir simulation with non-square grid blocks and anisotropic permeability. Soc Pet Eng J June: 531–543

Peaceman DW (1991) Presentation of a horizontal well in numerical reservoir simulation. SPE 21217, 11th SPE Symposium on Reservoir Simulation in Ananheim, California, Feb 17–20

Pinault JL, Schomburgk S (2006) Inverse modeling for characterizing surface water/groundwater exchanges. Water Resour Res 42:W08414. https://doi.org/10.1029/2005WR004587

Poeter EP, Hill MC (1997) Inverse models: a necessary next step in ground-water flow modeling. Ground Water 35:250–260

RamaRao BS, LaVenue AM, de Marsily G, Marietta MG (1995) Pilot point methodology for automated calibration of an ensemble of conditionally simulated transmissivity fields: 1. Theory and computational experiments. Water Resour Res 31(3):475–493

Rajanayaka C, Kulasiri D (2001) Investigation of a parameter estimation method for contaminant transport in aquifers. J Hydroinform 3:203–213

Sagar B, Yakowitz S, Duckstein L (1975) A direct method for the identification of the parameters of dynamic nonhomogeneous aquifers. Water Resour Res 11(4):563–570

Salo M (2008) Calderón problem. Lecture notes, Spring 2008 Mikko Salo, Department of Mathematics and Statistics, University of Helsinki. http://users.jyu.fi/~salomi/lecturenotes/calderon_lectures.pdf

Strack ODL (1989) Groundwater mechanics. Prentice Hall, Inc., Englewood Cliffs, New Jersey

Sun NZ (2004) Inverse problems in groundwater modeling. Kluwer Academic Publishers, Dordrecht

Tamburi A, Roeper U, Wexler A (1988) An application of impedance-computed tomography to subsurface imaging of pollution plumes. In: Collins AG, Johnson AI (eds) Ground-Water Contamination: Field Methods, ASTM Special Technical Publication 963, Americal Society for Testing and Materials: 86–100

Tarantola A (2005) Inverse problem theory and methods for model parameter estimation. Society of Industrial and Applied Mathematics (SIAM). ISBN: 978-0-89871-572-9

Tikhonov AN (1963a) Solution of incorrectly formulated problems and the regularization method. Sov Math Dokl 4:1035–1038

Tikhonov AN (1963b) Regularization of incorrectly posed problems. Sov Math Dokl 4:1624–1637

Tóth J (2009) Gravitational systems of groundwater flow. Cambridge University Press, Cambridge

Trykozko A, Zijl W, Bossavit A (2001) Nodal and mixed finite elements for the numerical homogenization of 3D permeability. Comput Geosci 5:61–64

Trykozko A, Brouwer GK, Zijl W (2008) Downscaling: a complement to homogenization. Int J Num Anal Model 5:157–70

Trykozko A, Mohammed GA, Zijl W (2009) Downscaling: the inverse of upscaling. In Conference on Mathematical and Computational Issues in the Geosciences. SIAM GS 2009, June 15–18, Leipzig

Valstar JR (2001) Inverse modeling of groundwater flow and transport. Ph.D. thesis Delft University of Technology, Delft, Netherlands

Valstar JR, McLaughlin DB, te Stroet CBM, van Geer FC (2004) A representer-based inverse method for groundwater flow and transport applications. Water Resour Res 40:W05116. https://doi.org/10.1029/2003WR002922

Warren JE, Price HS (1961) Flow in heterogeneous porous media. Soc Petrol Eng J 1(3):153–169. https://doi.org/10.2118/1579-G

Webster JG (1990) Electrical impedance tomography. Adam Hilger, Bristol

Wen XH, Deutsch CV, Cullick AS (1998) High resolution reservoir models integrating multiple-well production data. Soc Petrol Eng J 3(4):344–355

Wen XH, Deutsch CV, Gomez-Hernandez JJ, Cullick AS (1999) A program to create permeability fields that honor single-phase flow rate and pressure data. Comput Geosci 25(3):217–230

Wexler A (1988) Electrical impedivity imaging in two and three dimensions. Clin Phys Physio Meas 9, Suppl A:29–33

Wexler A, Fry B, Neuman MR (1985) Impedivity computed tomography algorithm and system. Applied Optics 24(23):3985–3992

Yeh WWG (1986) Review of parameter identification procedures in groundwater hydrology: the inverse problem. Water Resour Res 22(2):95–108

Yeh TC, Liu S (2000) Hydraulic tomography: development of a new aquifer test method. Water Resour Res 36(8):2095–2105

Yeh WWG, Yoon YS (1981) Aquifer parameter identification with optimum dimension in parameterization. Water Resour Res 17(3):664–672

Yorkey TJ, Webster JG (1987) A comparison of impedivity topographic reconstruction algorithms. Clin Phys Physiol Meas 8:55–62

Yorkey TJ, Webster JG, Tompkins WJ (1987) Comparing reconstruction algorithms for electrical impedance tomography. IEEE Trans Biomed Eng 34:843–852

Zhou H, Gómez-Hernández JJ, Li L (2014) Inverse methods in hydrogeology: evolution and recent trends. Adv Water Resour 63:22–37

Zijl W, Trykozko A (2001) Numerical homogenization of the absolute permeability using the conformal-nodal and mixed-hybrid finite element method. Transp Porous Med 44:33–62

Zimmerman DA, de Marsily GD, Gotway CA, Marietta MG, Axness CL, Beauheim RI, Bras RI, Carrera J, Dagan G, Davies PB (1998) A comparison of seven geostatistically based inverse approaches to estimate transmissivities for modeling advective transport by groundwater flow. Water Resour Res 34(6):1373–1413

Chapter 2
Foundations of Forward and Inverse Groundwater Flow Models

2.1 Basic Equations and Parameters

The equations governing groundwater flow through a porous medium are the water balance equation

$$s\frac{\partial h}{\partial t} + \nabla \cdot \mathbf{q} = 0 \qquad (2.1)$$

and Darcy's law

$$\mathbf{q} = -\underline{\mathbf{k}} \cdot \nabla h \qquad (2.2)$$

where s [L^{-1}] is the specific storage, h [L] is the head, $\mathbf{q}$ [L T^{-1}] is the flux density or specific discharge, $\underline{\mathbf{k}}$ [L T^{-1}] is the hydraulic conductivity tensor (in short the conductivity), t [T] is the time, ∇ [L^{-1}] is the gradient operator, [∇h is the vector with Cartesian components $(\partial h/\partial x, \partial h/\partial y, \partial h/\partial z)$], and $\nabla \cdot$ [L^{-1}] is the divergence operator ($\nabla \cdot \mathbf{q}$ is the scalar $\partial q_x/\partial x + \partial q_y/\partial y + \partial q_z/\partial z$). Equation 2.2 holds under conditions where the water density may be considered as constant. In Eq. 2.1, the specific storage term $s\,\partial h/\partial t$ accounts for the compressibility of the water and the pore space. To be precise, the terms flux and flux density have a different meaning. The flux density $q_n = \mathbf{q} \cdot \mathbf{n}$ [L T^{-1}] through a face with unit vector $\mathbf{n}$ [−] and surface area A [L^2] is equal to the flux Q [L^3 T^{-1}] through that surface area divided by the surface area of that face. However, in many texts and also in this book, the terms flux and flux density are used as synonyms, except when the difference is specified explicitly. In the literature, also the terms specific flux and specific discharge are used for flux density.

For incompressible flow, i.e., flow where compressible storage is negligibly small, the water balance (Eq. 2.1) simplifies to the continuity equation

© The Author(s) 2018
W. Zijl et al., *The Double Constraint Inversion Methodology*,
SpringerBriefs in Applied Sciences and Technology,
https://doi.org/10.1007/978-3-319-71342-7_2

$$\nabla \cdot \mathbf{q} = 0 \tag{2.3}$$

In many formulations, the right-hand sides of Eqs. 2.1 and 2.3 contain a source/sink term of the form $\sum_{n=1}^{N} Q_n \delta(\mathbf{x} - \mathbf{x}_n)$ to account for wells with injection rate Q_n or production rate $-Q_n$ [$L^3 T^{-1}$] in the points $\mathbf{x}_n$, where $\delta(\cdot)$ represents the Dirac delta function (Butkov 1973: 223–232; Morse and Feshbach 1953: 825).

In practical modeling applications, this source/sink approach is much simpler to apply than the more exact formulation based on well boundary conditions; in the limit of small well diameters, the two formulations are mathematically equivalent (Chavent and Jaffré 1986: 70–88). However, in applications where instead of the well rate the head is imposed, a source/sink approach is not applicable; in that case, the original equations (Eqs. 2.1 and 2.3) have to be applied. The well head has then to be imposed at the boundary of the well. In discretized models (numerical models), the head in the grid volume can be imposed. For nonzero production rates, the volume-centered head may differ from the measured well head; therefore, a near-well model (generally a relatively simple algebraic model) has to be used to relate the two heads (Peaceman 1977a, 1983, 1991; Chen and Zhang 2009).

Substitution of Darcy's law (Eq. 2.2) into the water balance equations (Eqs. 2.1 and 2.3) results in the groundwater flow equations for, respectively, compressible and incompressible flow

$$\begin{array}{cc}\text{Compressible flow} & \text{Incompressible flow} \\ s\frac{\partial h}{\partial t} = \nabla \cdot (\underline{\mathbf{k}} \cdot \nabla h) & \nabla \cdot (\underline{\mathbf{k}} \cdot \nabla h) = 0\end{array} \tag{2.4}$$

When there is nonzero groundwater flow ($\mathbf{q} \neq \mathbf{0}, \mathbf{e} = -\nabla h \neq \mathbf{0}$), the entropy per unit volume is increasing in time, which means that the dissipation rate $\rho g \mathbf{e} \cdot \underline{\mathbf{k}} \cdot \mathbf{e} > 0$ (heat produced per unit volume and unit time) is positive. As a consequence, hydraulic conductivity tensor $\underline{\mathbf{k}}$ is positive definite and specific storage s is non-negative. The conductivity tensor has the nine components, $k_{11}, k_{12}, k_{13}, k_{21}, k_{22}, k_{23}, k_{31}, k_{32}, k_{33}$. On the scale of a representative elementary volume, this tensor is symmetric (Olmstead 1968), i.e., $k_{12} = k_{21}, k_{23} = k_{32}, k_{31} = k_{13}$. Also, on coarser scales, this tensor may often be assumed to be symmetric; symmetry is exactly the case for periodic porous media (Trykozko et al. 2001), but is not generally the case (Zijl and Nawalany 1993). In an orthogonal curvilinear coordinate system, this positive-definite symmetric matrix has three orthogonal eigendirections with non-negative eigenvalues, or principal conductivities, $k_1 = k_{11} \geq 0, k_2 = k_{22} \geq 0, k_3 = k_{33} \geq 0$, while the cross terms vanish, i.e., $k_{12} = k_{23} = k_{31} = 0$. From now on, we assume that we know the principal directions of the conductivity tensor already before calibration. A popular choice is to apply a Cartesian coordinate system in which x and y represent the two horizontal directions and z the vertical direction and to assume $k_1 = k_x, k_2 = k_y, k_3 = k_z$.

The water table, or phreatic surface, forms the top boundary of the groundwater domain. The boundary conditions on the top boundary of the model play a prominent role in the so-called flow systems analysis (Bresciani et al. 2016;

El-Rawy et al. 2016; Tóth 2009). Flow systems analysis is one of the approaches that have inspired our approach to inverse modeling. Because of its importance, we will first give a rigorous description of the exact water table conditions (Eqs. 2.5 and 2.6). Then, we present a simpler and perhaps more insightful approximation (given by Eq. 2.7).

Denoting the water table height above a horizontal reference level as $z = \zeta(x, y, t)$, the kinematics of the water table (its movements in time) are governed by the kinematic condition

$$\theta \frac{\partial \zeta}{\partial t} + q_x \frac{\partial \zeta}{\partial x} + q_y \frac{\partial \zeta}{\partial y} - q_z = r_{eff} \text{ on } z = \zeta(x, y, t) \qquad (2.5)$$

where $\theta(x, y, \zeta)$ [−] is the specific yield, or effective porosity, and $r_{eff}(x, y, t)$ [L T^{-1}] is the recharge rate, or effective precipitation, arriving at the water table from the unsaturated zone above the water table. In flux models, i.e., in models where the flux is imposed on the boundaries, the recharge rate r_{eff} is imposed. (To be precise, in Eq. 2.5 it is assumed that the recharge flux density, r_{eff}, is vertical. Its component normal to the curved phreatic surface is $r_{eff}/[1 + (\partial \zeta/\partial x)^2 + (\partial \zeta/\partial y)^2]^{1/2}$.

The dynamics of the water table, i.e., the forces exerted in the water table, are governed by the fact that at the water table the water pressure is equal to the atmospheric pressure. This is equivalent to the statement that, on the water table, the head, h, is equal to the water table height, ζ, resulting in the dynamic boundary condition

$$h(x, y, \zeta(x, y, t), t) = \zeta(x, y, t) \text{ on } z = \zeta(x, y, t) \qquad (2.6)$$

In "head models," i.e., in models where the head is imposed on the boundaries, head h is imposed on the water table. For instance, flow systems analysis (Tóth 2009) is based on such a head model. From the dynamic condition (2.6), it follows that $\partial h/\partial \lambda = (\partial \zeta/\partial \lambda)/(1 - \partial h/\partial z)$, $\lambda = x, y, t$, and substitution into kinematic condition (2.5) leads to an expression for the recharge rate r_{eff} as a function of the head and its derivatives with respect the horizontal coordinates x and y, vertical coordinate z, and time t (Bear 1972). For a detailed explanation of the physics underlying the water table equations (Eqs. 2.5 and 2.6), see Bear (1972), Zijl and Nawalany (1993), and De Smedt and Zijl (2018); these books present also a derivation of the well-known two-dimensional Dupuit–Forchheimer equation for flow in unconfined aquifers (water-bearing ground layers in which the periodically rising and falling water table resides).

To illustrate the essence of the above-presented water table description, we present a rough approximation to water table equations (Eqs. 2.5 and 2.6). Assuming that $|\partial h/\partial z| \ll 1$, we may replace in Eq. 2.5 water table height function $\zeta(x, y, t)$ with head function $h(x, y, z, t)$. This assumption can be justified in many practical applications, and the often-applied hydraulic approximation, or Dupuit–Forchheimer approximation, is based on this assumption (De Smedt and Zijl 2018). However, for the purpose of demonstration, we also assume that in

Eq. 2.5 the terms $q_x \partial h/\partial x$ and $q_y \partial h/\partial y$ are negligibly small with respect to q_z. This linearization leads to the simple (often too simple), but insightful expression

$$-q_z = r_{eff} - \theta \frac{\partial h}{\partial t} \quad \text{on the top boundary} \qquad (2.7)$$

Since the vertical direction is defined positive upward, the term $-q_z$ in Eq. 2.7 is the downward directed vertical groundwater flux density. Equation 2.7 illustrates clearly the water balance on the water table. On the top boundary of the model domain, i.e., on the water table $h = z$, the downward directed vertical groundwater flux, $-q_z$, is equal to the recharge rate, r_{eff}, minus the increase in the amount of groundwater, $\theta \partial \zeta/\partial t$, flowing into the model domain. In a time interval Δt, the volume of the model domain is increased by the raised water table height $\Delta \zeta = (\partial h/\partial t)\Delta t$. Although the phreatic boundary conditions have to be applied at the phreatic surface $z = \zeta(x, y, t)$, the linearized condition (Eq. 2.7) may sometimes be applied on a fixed horizontal top boundary $z = h_0$ representing the water table height averaged over space and time. This geometric linearization can be justified for cases in which the variations and fluctuations of the water table height are relatively small with respect to the depth of the basin under consideration. Combined with Eq. 2.7, this results in a linear groundwater flow problem that can be solved by analytical methods; see, for instance, Tóth (2009: 41–46), El-Rawy et al. (2015).

In conventional "flux models," i.e., in forward models where the recharge rate r_{eff}^F is imposed on the top boundary, the solution of groundwater flow Eq. 2.4 yields head h^F and groundwater flux $\mathbf{q}^F$ in the model domain as well as on the boundaries including the top boundary (superscript F refers to the flux model). On the other hand, in Tóthian "head models," i.e., in forward models where the head, h^H, is imposed on the top boundary, the solution of groundwater flow Eq. 2.4 yields the head h^H and groundwater flux $\mathbf{q}^H$ in the model domain as well as on the boundaries including the top boundary (superscript H refers to the head model). Provided that specific yield θ is known, Eq. 2.7 then results in the recharge rate r_{eff}^H. The requirement that recharge rate r_{eff}^F imposed on the flux model matches with recharge rate r_{eff}^H obtained from the head model can be used to modify the original "old" conductivity k_z^{old} in a corrected "new" conductivity k_z^{new} by the simple correction equation $k_z^{new} = -q_z^F/(\partial h^H/\partial z) = k_z^{old}(q_z^F/q_z^H)$ based on Darcy's law. This idea has been worked out and justified in more detail in Chaps. 3 and 6, while the case studies presented in Chaps. 5 and 6—in which both flux and head models are applied in coupled mode—have generally been based on the linearized approximation (Eq. 2.7 applied on an averaged top boundary).

The four parameters, k_x, k_y, k_z, and s, are assumed to be dependent on position x, y, z, but independent from time t (as has been justified above, we neglect the six cross terms k_{xy}, k_{yz}, k_{zx}, k_{yx}, k_{zy}, k_{xz} of the conductivity tensor). The fifth parameter, the specific yield θ, is only dependent on the horizontal coordinates x and y. It is generally assumed that this parameter is time-independent, but it dependent on the timescale of the process under consideration. On timescales where $|\partial h/\partial t|$ is relatively large, θ is small, while on timescales where $|\partial h/\partial t|$ is relatively small, θ is relatively large. If the flow is incompressible (negligible term $s \, \partial h/\partial t$ in Eq. 2.1) and if in addition the phreatic storage term $s \, \theta \, \partial \zeta/\partial t$ in Eq. 2.5 is negligible, the flow is called quasi-steady (also see Chap. 4, Sect. 4.1 and Chap. 6, Sect. 6.4). Although we will pay some attention to the two storage parameters s and θ in Chap. 4, Sect. 4.1, this book focuses mainly on the hydraulic conductivity.

2.2 Introduction to Calderón's Conjecture

In a highly cited paper (1215 cites Google Scholar), Calderón conjectured that, under some smoothness conditions, the conductivity field in a model domain can be determined uniquely from the head and the normal flux imposed on the closed boundary of that model domain (Calderón 1980). To be more specific, Calderón considered equation $\nabla \cdot k \, \nabla h = 0$ for incompressible flow (see Eq. 2.4). Calderón used this equation in the context of electrical impedance tomography for geophysical and medical imaging. He considered an electricity conducting body with isotropic conductivity k and electric potential h where the potential and the normal current density (normal flux) are imposed on the boundary; see, for instance, Cheney et al. (1999), Borcea (2002, 2003), Borcea et al. (2003), Holder (2005). To be sure, in hydrogeology we cannot know both head and normal flux on the boundary; head–flux pairs are generally known only at some places on the external boundary and/or internal boundaries (wells, including monitoring wells). However, because Calderón's conjecture is inspiring for the inverse methods presented in the next chapters focusing on hydrogeological problems, we will present below an argument for the validity of Calderón's conjecture.

We consider an isotropic porous medium with heterogeneous conductivity $k = \alpha^2 \geq 0$. To avoid the occurrence of negative conductivities in the conductivity estimation procedure, we base the theories presented in this book as much as possible on the square root of the conductivity, in short denoted as the sqrt-conductivity $\alpha = \pm k^{1/2}$. The heterogeneous sqrt-conductivity may be discontinuous. However, we also pay some attention to subsurface bodies in which the sqrt-conductivity is a smooth function of space, as may, for instance, be the case within aquifers or aquitards.

Under the condition that the heterogeneous square root conductivity $\alpha = \alpha \, (x, y, z)$ has at least continuous second derivatives, it follows from straightforward mathematical manipulations that the groundwater flow equation (Eq. 2.4) is mathematically equivalent with the equation $\nabla^2 h\alpha - h\nabla^2\alpha = (s/\alpha) \, \partial h/\partial t$, which can be written in two coupled equations for the three functions α $[\mathrm{L}^{-1/2} \, \mathrm{T}^{-1/2}]$, $\omega = h\alpha$ $[\mathrm{L}^{-3/2} \, \mathrm{T}^{-1/2}]$, and $\tau = \alpha^{-1}\nabla^2\alpha$

$$\nabla^2 \alpha - \tau\alpha = 0 \qquad (2.8)$$

$$\nabla^2 \omega - \tau\omega = \frac{s}{\alpha^2} \frac{\partial \omega}{\partial t} \qquad (2.9)$$

In the literature on Calderón's conjecture, the analysis is generally based on Eqs. 2.8 and 2.9 (Salo 2008). Following their approach, we have to invoke the condition that $\alpha = \alpha(\mathbf{x})$ is a smooth function of position vector $\mathbf{x}$ with Cartesian components x, y, z. In realistic hydrogeological applications, the hydraulic conductivity is generally discontinuous. At faces of discontinuous conductivities, the head, h, and the normal component of the flux density, $q_n = -k \,\partial h/\partial n$, are continuous. As a consequence, the normal derivative of sqrt-conductivity α is discontinuous, which means that derivatives of α do not exist at discontinuity faces. In such cases, Eqs. 2.8 and 2.9 are not at all equivalent to the groundwater flow equation (Eq. 2.4).

However, within certain subsurface units, it is often reasonable to assume spatially correlated conductivities. For instance, zonation, in which the conductivity is assumed to be homogeneous within a number of zones, is a technique for inverse modeling (see Chap. 6). A discontinuous conductivity field cannot be fully determined from measured head and flux data on the boundaries of a model domain (wells, including monitoring wells, may be considered as internal boundaries); for a further analysis of this case, see Sect. 2.4.

Under the assumption that the conductivity field is smooth (differentiable), we are able to add a third equation to the two coupled equations (Eqs. 2.8 and 2.9); for a further analysis of this case, see Sect. 2.4. To show the way how this can be accomplished, we expand sqrt-conductivity field $\alpha(\mathbf{x})$ in a series of meaningfully chosen basis functions, or representer functions, $u_i(\mathbf{x})$; i.e.,

$$\alpha(\mathbf{x}) = \sum_{i=1}^{m} \gamma_i u_i(\mathbf{x}) \qquad (2.10)$$

where γ_i is a coefficient. The m coefficients $\gamma_1, \gamma_2, \ldots, \gamma_m$ are related to the values $\alpha_1, \alpha_2, \ldots, \alpha_m$ of the sqrt-conductivity in m points $\mathbf{x}_1, \mathbf{x}_2, \ldots, \mathbf{x}_m$ on the boundaries. This results in a linear relation between $\gamma_1, \gamma_2, \ldots, \gamma_m$ and $\alpha_1, \alpha_2, \ldots, \alpha_m$.

$$\alpha_j = \alpha(\mathbf{x}_j) = \sum_{i=1}^{m} \gamma_i u_i(\mathbf{x}_j) \qquad (2.11)$$

In this way, the sqrt-conductivities α within the model domain are linearly related to the m sqrt-conductivities α_j on the boundaries. Meaningful basis functions for this approach may be exponential functions (Uhlmann 2003, 2009; Salo 2008) or powers of x, y, and z; see Sect. 2.4 for a detailed analysis.

Substitution of Eq. 2.10 into Eq. 2.8 yields $\tau = \left(\sum_{i=1}^{m} \alpha_i \nabla^2 u_i\right) / \left(\sum_{i=1}^{m} \alpha_i u_i\right)$, and considering incompressible flow (flow for which $(s/\alpha^2)\, \partial\omega/\partial t$ in Eq. 2.4 is negligible), substitution into Eq. 2.9 results in an equation for ω

$$\nabla^2 \omega - \frac{\sum_{i=1}^{m} \gamma_i \nabla^2 u_i}{\sum_{i=1}^{m} \gamma_i u_i} \omega = 0 \text{ in the model domain} \qquad (2.12)$$

For given α_i and ω_i on the boundary, Eq. 2.12 (or its discretized approximation) can be solved to determine ω in the whole model domain, from which also the head $h = \omega/\alpha$ is known in the whole model domain. However, instead of specifying m values of sqrt-conductivity α_i on the boundary, we may impose m normal flux values $q_{m;i} = -\alpha_i^2 \partial h/\partial n$ on the boundary. Doing so, Eq. 2.12, combined with this flux boundary condition and this head boundary condition, has been turned into a nonlinear problem to determine the m sqrt-conductivities on the boundary, as well as the heads and the sqrt-conductivities in the model domain. This problem could then be solved iteratively (also see Chap. 2, Sect. 2.4). The above-presented argument shows a way how to tackle Calderón's conjecture (in which $m \rightarrow \infty$). For more details, see Sect. 2.4. The solution exists only if the boundary conditions allow for a nonnegative conductivity field. This latter requirement is not always satisfied in hydrology, where the imposed boundary conditions may be based on too inaccurately measured data to allow for a nonnegative solution.

The above-presented original Calderón problem is inspiring, because it shows that for inverse modeling (parameter estimation) more heads and fluxes have to be imposed than required for forward modeling (specified parameters). However, in realistic groundwater flow problems, we do not know head and flux on the full boundary; we know head and flux only in a limited number of observation wells (zero flux, head measured) and, sometimes, on the water table (estimated water table height and recharge rates). Therefore, we will from now on focus on the incomplete Calderón problem, in which only a limited number of flux and head data are measured, often with rather great inaccuracy. Fluxes are imposed only at locations where fluxes are known, while heads are imposed only at locations where heads are known. In this incomplete Calderón problem, the conductivity field is not unique, but depends on the initial choice of the conductivity field. Moreover, because in hydrogeology the conductivity is generally discontinuous, $\nabla^2 \alpha$ does not exist, and therefore, Eqs. 2.8, 2.9, and 2.10 cannot be applied. For that reason, we introduce in Chap. 3 a "Calderón-inspired" approach to parameter estimation that is directly based on the water balance equation (Eq. 2.1) and, to avoid negative conductivities, on Darcy's law (Eq. 2.2) formulated with sqrt-conductivity α, as well as on the imposed flux and head conditions.

2.3   Stefanescu's Alpha Centers

Stefanescu (1950) introduced the so-called alpha center method in the context of geo-electrical problems. His method is based on solutions of Eqs. 2.8 and 2.9 with $\partial \omega / \partial t = 0$. In his approach $\tau = 0$ except in a number of singular points (x_n, y_n, z_n), $n = 1, 2, \ldots, N$, the so-called alpha centers. The components α_n of the sqrt-conductivity field $\alpha = \sum_{n=1}^{N} \alpha_n = \pm k^{1/2}$ are then given as

$$\alpha_n = \frac{A_n}{\sqrt{(x - x_n)^2 + (y - y_n)^2 + (z - z_n)^2}} \tag{2.13}$$

From Eq. 2.8, it follows then that, in the limit $(x, y, z) \rightarrow (x_n, y_n, z_n)$, τ_n is equal to

$$\tau_n = -4\pi \sqrt{(x - x_n)^2 + (y - y_n)^2 + (z - z_n)^2} \times \delta(x - x_n)\delta(y - y_n)\delta(z - z_n) \tag{2.14}$$

where $\delta(\cdot)$ represents the Dirac delta function (Butkov 1973: 223–232; Morse and Feshbach 1953: 825). From the definition of the Dirac delta function, it follows that $\tau_n = 0$ and, as a consequence, $\tau = \sum_{n=1}^{N} \tau_n = 0$ outside the alpha centers (x_n, y_n, z_n) and may have any value (is undetermined) in the alpha centers.

Since Eq. 2.9 with $\partial \omega / \partial t = 0$ has the same form as Eq. 2.8, a similar solution holds for ω_n

$$\omega_n = \frac{\Omega_n}{\sqrt{(x - x_n)^2 + (y - y_n)^2 + (z - z_n)^2}} \tag{2.15}$$

From Eqs. 2.14 and 2.15, it follows that the head is

$$h = \frac{\omega}{\alpha} = \frac{\sum_{n=1}^{N} \omega_n}{\sum_{n=1}^{N} \alpha_n} \tag{2.16}$$

In the alpha centers (x_n, y_n, z_n), the head is equal to the finite value $h_n = \Omega_n / A_n$; i.e., the head is a smooth function of x, y, z without singularities.

Finally, the $2N$ constants A_n, Ω_n can be adjusted to honor both the head boundary condition and the flux boundary condition at the N alpha centers. This approximation method shows that inverse problems can be solved approximately using the two coupled equations (Eqs. 2.8 and 2.9) with both head and flux boundary conditions.

Stefanescu's approach is inspiring because of its assumption $\nabla^2 \alpha = 0$. Although not directly applicable to groundwater modeling, this idea can be generalized to assuming $\nabla^{2m} \alpha = 0$ for the construction of basis functions $u_i(x, y, z)$ and coefficients γ_i in which sqrt-conductivity $\alpha(x, y, z)$ is expanded to solve the Calderón problem; see Sect. 2.4.

2.4   In Depth: A Closer Look at Calderón's Conjecture

In this section, we analyze the so-called "Calderón problem" in more detail. Calderón conjectured that from the potential equation $\nabla \cdot k \nabla h = 0$ (Eq. 2.4) with "sufficiently smooth" nonnegative conductivity field $k(x, y, z) \geq 0$, the conductivity can be determined uniquely from exact measurements of head h and normal flux $q_n = -k\, \partial h / \partial n$ on the closed boundary of the model domain. For inaccurately measured conditions, a conductivity that is nonnegative everywhere in the domain cannot always be found. The Calderón problem is then to determine what "sufficiently smooth" means to guarantee existence and uniqueness, and to find solutions.

For the estimation of conductivities in grid volumes, we apply Calderón's idea of "conductivity smoothness" to correlate the conductivities in neighboring grid blocks. In this framework, existence and uniqueness need not be considered: A "sufficiently accurate" approximate solution can always be found.

In Sect. 2.4.1, we investigate how many uncorrelated conductivities (in the grid volumes of a numerical model) can be derived from specification of a finite number of imposed flux and head conditions, and in the remaining sections (Sects. 2.4.2, 2.4.3, 2.4.4, 2.4.5, and 2.4.6), we show how we can determine many more conductivities if we correlate them by "smoothness conditions."

2.4.1   Uncorrelated Conductivity Fields

In this subsection, we investigate the following question: How many conductivities can be determined from $N + 1$ measured flux–head pairs?

Let us consider a groundwater flow domain with $N + 1$ nodes with node numbers $n = 1, 2, \ldots, N + 1$, where we know both the flux of water Q_n [$L^3\, T^{-1}$] into each node n and the head h_n [L^3] in each node n. Only the differences in measured head play a role in groundwater flow (see Eq. 2.2), which means that all heads could be measured with respect to head h_{N+1}. That is, only the N measured head differences $h_n - h_{N+1}$, $n = 1, 2, \ldots, N$, play a role. Also, for incompressible flow, only N fluxes are measured; the $N + 1$st flux follows from continuity equation $\sum_{n=1}^{N+1} Q_n = 0$ (flow in = flow out; also see Eq. 2.3).

From the point of view of the $N + 1$ nodes, the subsurface is "seen" as a network of $N(N + 1)/2$ conductors mutually connecting the $N + 1$ nodes. Each conductor has a conductance defined as flux through the conductor divided by head difference between the conductor's end nodes. As will be shown below, only N conductivities can be determined from the N measured fluxes and head differences, the other $N(N - 1)/2$ conductivities have to be obtained from additional knowledge (the soft data).

Let us consider nodes 1 and 2 with measured heads h_1 and h_2, respectively. Head difference $h_1 - h_2$ causes flow through the conductor connecting node 1 to node 2. If we know flux $Q_{1 \to 2}$ from node 1 to node 2, we can determine conductance $K_{1 \leftrightarrow 2} = Q_{1 \to 2}/(h_1 - h_2)$ [$L^2\, T^{-1}$] (see Darcy's law, Eq. 2.2). To determine flux $Q_{1 \to 2}$, we assign realistically chosen values (the soft data) to the remaining $N - 1$ conductances $K_{1 \leftrightarrow n}$, $n = 3, 4, \ldots, N+1$, connecting node n to node 1. From the measured head differences between all nodes, we can then determine the $N - 1$ fluxes $Q_{1 \to n} = K_{1 \leftrightarrow n}(h_1 - h_n)$. The measured total flux of water flowing into node 1 is equal to $Q_1 = Q_{1 \to 2} + \sum_{n=3}^{N+1} Q_{1 \to n}$, from which we determine $Q_{1 \to 2}$ resulting in $K_{1 \leftrightarrow 2}$.

A similar procedure can be set up for node 2. Because we know already conductance $K_{1 \leftrightarrow 2}$, we have to assign $N - 2$ realistically chosen values (soft data) of the conductances $K_{2 \leftrightarrow n}$, $n = 4, 5, \ldots, N+1$, to the $N - 2$ remaining conductors connecting node n to node 2; only conductance $K_{2 \leftrightarrow 3}$ is left unspecified. From these conductances and from the known head differences between their nodes, we calculate the $n - 2$ fluxes $Q_{2 \to n} = K_{2 \leftrightarrow n}(h_2 - h_n)$. The measured total flux of water flowing into node 2 is equal to $Q_2 = Q_{1 \to 2} + Q_{2 \to 3} + \sum_{n=4}^{N+1} Q_{2 \to n}$, from which we determine $Q_{2 \to 3}$ resulting in $K_{2 \leftrightarrow 3} = Q_{2 \to 3}/(h_2 - h_3)$.

Continuation of this procedure shows that, in the end, we have to assign values (soft data) to $N(N - 1)/2$ conductances to determine N conductances from the measured N flux–head pairs in the nodes.

Now, we consider a gridded numerical model. The above-presented "tubes with conductances" are now made up of the grid block conductivities in the model. Also in this case holds that from $N + 1$ measured heads (i.e., N measured head differences) and N measured fluxes, the conductivities of only N grid volumes can be determined. In fact, we know more, because the soft data, i.e., the chosen $N(N - 1)/2$ conductivity values, cannot be chosen arbitrarily; they have to be chosen in such a way that the calculated N conductivities are nonnegative. Smoothing, i.e., allowing only small differences between conductivity values in neighboring grid blocks, may help to avoid negative conductivities.

The above-presented conductance network procedure is an example of direct inversion as defined by Yeh (1986), from which we observe that small head gradients, as they often occur in flat deltaic regions, makes the solution unstable because of the divisions by $h_n - h_{n*}$ with $h_n \approx h_{n*}$. Smoothing will stabilize the inversion procedure (also see the discussion on Chap. 3, Sect. 3.9.2.2). For the above-mentioned reasons, Sect. 2.4.2 explores the concept of smoothness in more detail.

2.4.2   Smoothed Conductivity Fields

Defining sqrt-conductivity $\alpha = \pm k^{1/2}$ and assuming that α is sufficiently smooth to allow for the existence of $\nabla^2 \alpha$, the groundwater flow equation (Eq. 2.4) can be written as Eqs. 2.8 and 2.9. Elimination of τ from these equations results in

$$\nabla^2 h\alpha - h\nabla^2\alpha = (s/\alpha)\,\partial h/\partial t \tag{2.17}$$

Terms like "smooth" and "sufficiently smooth" to denote some form of differentiability, as well as the mathematics to handle such notions, are used here in an intuitive way based on quantized (discretized) space. The mathematical framework in which these terms are rigorously defined and applied for continuous space can, for instance, be found in Duvaut and Lions (1976).

In a differentiable conductivity field, neighboring conductivities are related to each other by the well-known Taylor series expansion

$$\begin{aligned}
\alpha(x+\Delta x) = {} & \alpha(x) + (\partial\alpha/\partial x)_x \Delta x \\
& + \frac{1}{2}(\partial^2\alpha/\partial x^2)_x \Delta x^2 + \cdots + \frac{1}{n!}(\partial^n\alpha/\partial x^n)_x \Delta x^n + \cdots
\end{aligned} \tag{2.18}$$

We exemplify our approach for a two-dimensional grid with discretization lengths $\Delta x = \Delta y = \Delta$. Then, the finite difference approximations of the terms in Eq. 2.17 are

$$\begin{aligned}
(\nabla^2 h\alpha)_{i,j} \approx {} & [(h\alpha)_{i-1,j} + (h\alpha)_{i+1,j} \\
& + (h\alpha)_{i,j-1} + (h\alpha)_{i,j+1} - 4(h\alpha)_{i,j}]/\Delta^2 \\
h_{i,j}(\nabla^2\alpha)_{i,j} \approx {} & h_{i,j}(\alpha_{i-1,j} + \alpha_{i+1,j} \\
& + \alpha_{i,j-1} + \alpha_{i,j+1} - 4\alpha_{i,j})/\Delta^2
\end{aligned} \tag{2.19}$$

and substitution of Eq. 2.19 into Eq. 2.17 yields

$$\begin{aligned}
& \alpha_{i+1,j}(h_{i,j} - h_{i+1,j}) - \alpha_{i-1,j}(h_{i-1,j} - h_{i,j}) \\
& + \alpha_{i,j+1}(h_{i,j} - h_{i,j+1}) - \alpha_{i,j-1}(h_{i,j-1} - h_{i,j}) \\
& = -(s/\alpha)\partial h/\partial t
\end{aligned} \tag{2.20}$$

From now on, we simplify our presentation by considering incompressible flow (negligible term $s\,\partial h/\partial t$); generalization to compressible flow is straightforward. In the block-centered finite difference method, Darcy's law for the flux in the x-direction through the face between grid block $i+1, j$ and grid block i,j is approximated by

$$q_{x,\,i+\frac{1}{2},j} = -\alpha^2_{i+\frac{1}{2},j}(h_{i,j} - h_{i+1,j}) \tag{2.21}$$

where $\alpha^2_{i+\frac{1}{2},j}$, the effective conductivity at that face, is equal to the harmonic average of the conductivities in the grid blocks $i+1, j$ and i,j. Similar approximations follow for the other three faces. Substitution of Eq. 2.21 into Eq. 2.20 yields

$$\frac{\alpha_{i+1,j}}{\alpha_{i+\frac{1}{2},j}^2} q_{x,i+\frac{1}{2},j} - \frac{\alpha_{i-1,j}}{\alpha_{i-\frac{1}{2},j}^2} q_{x,i-\frac{1}{2},j}$$
$$+ \frac{\alpha_{i,j+1}}{\alpha_{i,j+\frac{1}{2}}^2} q_{y,i,j+\frac{1}{2}} - \frac{\alpha_{i,j-1}}{\alpha_{i,j-\frac{1}{2}}^2} q_{y,i,j-\frac{1}{2}} = 0 \tag{2.22}$$

Thanks to the water balance (Eq. 2.1), the fluxes through the faces are continuous, and in the model domain, the fluxes change only gradually. Equation 2.22 shows then that not only the fluxes, but also the conductivities vary smoothly in the model domain. This is clearly demonstrated for one-dimensional flow, where $q_{x,i+\frac{1}{2},j} = q_{x,i-\frac{1}{2},j} = q_0$ is constant. From Eq. 2.22, it follows then that the smoothest possible solution is $\alpha_{i+1,j} = \alpha_{i+\frac{1}{2}j} = \alpha_{i-1,j} = \alpha_0$ is constant. Of course, there are infinitely more smooth solutions, but here we consider the smoothest possible solution.

Since the flux runs and head runs with a numerical model (e.g., MODFLOW) for the double constraint method are based on the conventional formulation (Eq. 2.4), the resulting sqrt-conductivity field will not be smooth. To smooth the sqrt-conductivity, we propose an additional step in the double constraint iterations.

Equation 2.17 (for incompressible flow) can also be written as $\nabla^2 \alpha = (h + h_0)^{-1} \nabla^2 [(h + h_0) \alpha]$ in which h_0 is a suitably chosen reference head. To avoid division by zero, we choose $h_0 > -h_{min}$, where h_{min} is the smallest value of h in the model domain. Then, $\tilde{h} = h + h_0$ is positive everywhere in the model domain.

Now, we consider the original double constraint method, in which each grid volume (each "point") is a zone. Consider a sqrt-conductivity α obtained after an iteration step with head $\tilde{h}$ in the head run. From the α-values in the grid volume centers, we calculate $b = \tilde{h}^{-1} \nabla^2 \tilde{h} \alpha$ by numerical differentiation (using an expression like Eq. 2.19). According to Eq. 2.17, b may be considered as a source term for Poisson equation $\nabla^2 \alpha' = b$. Using the sqrt-conductivity obtained from the iteration step as boundary condition for this Poisson equation (i.e., applying Dirichlet boundary condition $\alpha'|_{\partial D} = \alpha|_{\partial D}$), the smoothed sqrt-conductivity α' is then obtained by solving the Poisson equation. This can simply be done by a groundwater flow model (e.g., MODFLOW). We use the thus-obtained smoothed α' as initial sqrt-conductivity for the next iteration step by the double constraint method. Repeating this procedure after every DCM iteration step will then result to converge, i.e., to $\alpha \to \alpha'$ in the grid block centers. Instead of smoothing the whole model domain, we could also select a number of zones in which we want to smooth.

2.4.3   Smoothest Possible Sqrt-Conductivity Fields

We consider a two- or three-dimensional domain D enclosed by boundary ∂D in which the conductivity is isotropic but heterogeneous; i.e., $k = k(x, y, z)$ is a function of the spatial coordinates x, y, z (for a one-dimensional domain problem, see Sect. 2.4.6). We require that this heterogeneous conductivity field is "the smoothest possible" conductivity field. That is, if everywhere on boundary ∂D the conductivity has the same value, k_B, independent from its spatial position, the conductivity in domain D is homogeneous and equal to its boundary value k_B. This requirement corresponds with Calderón's aim to develop a method that can detect an internal deviation from homogeneity by voltage and current measurements on the surface. In this way we can detect a well-conducting tumor in a poorly conducting breast, or a well-conducting salt lens in a poorly conducting fresh water body (Uhlmann 2003, 2009). For hydrogeological applications, voltage is replaced by head and current by flux, for instance, to detect a poorly conducting clay lens in a well-conducting aquifer.

In Stefanescu's approach (Chap. 2, Sect. 3), the construction of an inverse solution is based on the simple assumption $\tau = 0$ leading to $\nabla^2\alpha = 0$ and $\nabla^2 h\alpha = 0$ (see Eqs. 2.8 and 2.9). Although Stefanescu's approximation may lead to acceptable conductivity estimations in simple applications, application to more comprehensive problems has to be based on more general and flexible assumptions. Instead of Stefanescu's assumption $\nabla^2\alpha = 0$ (Sect. 2.3), we base the inverse solution of sqrt-conductivity function $\alpha(x, y, z)$ on equation $\nabla^{2(L+1)}\alpha = 0$. For a sufficiently large number L, the groundwater flow equation $\nabla^2\alpha = h^{-1}\nabla^2 h\alpha$, or its more usual mathematically equivalent $\nabla \cdot \alpha^2\nabla h = 0$, is then used to determine a sqrt-conductivity field α in domain D that approximately honors the head and flux conditions h_B and q_B on boundary ∂D. This approach implies that we have solved the equation $\nabla^{2L}(h^{-1}\nabla^2 h\alpha) = 0$ instead of Stefanescu's equation $\nabla^2 h\alpha = 0$.

In principle, the above-introduced approach can be used for model domains with an arbitrary shape. However, to illustrate the idea, we consider a rectangular or block-shaped domain D with closed boundary $0 \leq x \leq D_x$, $0 \leq y \leq D_y$, $0 \leq z \leq D_z$. For this domain, we construct a finite number of L polynomials $p_{L-\ell}$ ($\ell = 1, 2, \ldots, L < \infty$) in x, y, z, in such a way that the polynomials preserve their form when the coordinates are interchanged (e.g., when $x \to y$, $y \to z$, $z \to x$ or $x \to z$, $y \to x$, $z \to y$). In addition, we consider polynomials of finite degree.

To avoid nonzero solutions α for which $\alpha = 0$ on the boundary (spurious solutions), we derive conditions for function τ in the coupled equations $\nabla^2\alpha - \tau\alpha = 0$ and $\nabla^2\omega - \tau\omega = 0$ (Eqs. 2.8 and 2.9, respectively) for incompressible flow. Sqrt-conductivity α is expanded in a series of linearly independent basis functions, or representer functions, $u(x), v(y), w(z)$ in such a way that $\alpha = \sum_{ijk} \gamma_{ijk}\, u_i\, v_j\, w_k$, where the γ_{ijk}'s are the coefficients. For each component $\alpha_{ijk} = \gamma_{ijk}u_iv_jw_k$ in this series expansion, we find $\tau_{u_iv_jwk} = \alpha_{ijk}^{-1}\nabla^2\alpha_{ijk} = u_i^{-1}\partial^2 u_i/\partial x^2 + v_j^{-1}\partial^2 v_j/\partial y^2 +$. A boundary condition for α on the closed boundary ∂D (a Dirichlet condition for Eq. 2.8) leads to a unique solution only if at least one of the functions $\tau_{u_i} = u_i^{-1}\partial^2 u_i/\partial x^2$, $\tau_{v_j} = v_j^{-1}\partial^2 v_j/\partial y^2$, $\tau_{w_k} = w_k^{-1}\partial^2 w_k/\partial z^2$ is nonnegative. If all three functions $\tau_{u_i}, \tau_{v_j}, \tau_{w_k}$ derived from function $u_i(x)\, v_j(y)\, w_k(z)$ are negative, there exist spurious solutions, which means that the solution of $\nabla^2\alpha - \tau\alpha = 0$ is not unique. For example, for the above-defined block-shaped domain the solution $\alpha_{ijk} = \gamma_{ijk} \sin(\kappa_x x) \sin(\kappa_y y) \sin(\kappa_z z)$ results for $\kappa_x = \pm m\pi/D_x$, $\kappa_y = \pm m\pi/D_y$, $\kappa_z = \pm m\pi/D_z$ ($m = 1, 2, 3, \ldots$) in nonzero values of α inside domain D for zero boundary values $\alpha_B = 0$. In that case, $\tau_{u_i} = -\kappa_x^2 < 0$, $\tau_{v_j} = -\kappa_y^2 < 0$, and $\tau_{w_k} = -\kappa_z^2 < 0$ are negative. On the other hand, a solution like $\alpha_{ijk} = \gamma_{ijk} \sin(\kappa_x x) \sin(\kappa_y y) \exp(\kappa_z z)$ cannot give rise to nonzero solutions for the Dirichlet boundary conditions $\alpha_B = 0$. Although $\tau_{u_i} = -\kappa_x^2 < 0$ and $\tau_{v_j} = -\kappa_y^2 < 0$ are negative, spurious solutions do not occur because $\tau_{w_k} = \kappa_z^2 > 0$ is positive. Polynomial basis functions $u_i(x) = x^i$, $v_j(y) = y^j$, $w_k(z) = z^k$, $i, j, k = 0, 1, 2, 3, \ldots$, do not result in spurious solutions, as well as exponential basis functions [upon which "complex geometrical optics" is based (Salo 2008: 18)].

2.4.4 Polynomial Construction

To simplify the presentation, we consider a two-dimensional rectangular domain; the extension to the above-introduced three-dimensional rectangular domain is straightforward.

For the ease of presentation, we assume that the domain is a square with center $(x_1, y_1) = (\delta, \delta)$, where $2\delta = D_x = D_y = D_z$. The polynomial solution of Poisson equation $\nabla^2 p_L = c_L$, where c_L is a constant, is $p_L = (a_x x^2 + b_x x) + (a_y y^2 + b_y y) + c$, from which it follows that $c_L = \nabla^2 p_L = 2(a_x + a_y)$. Specifying the four boundary conditions, the four polynomial values $p_{L;0,1}, p_{L;2,1}, p_{L;1,0}, p_{L;1,2}$ result in the following four equations for the four polynomial constants: $a_y \delta^2 + b_y \delta = p'_{L;0,1}$, $a_x \delta^2 + b_x \delta = p'_{L;1,0}$, $4a_x \delta^2 + 2a_x \delta + a_y \delta^2 + b_x \delta = p'_{L;2,1}$, and $a_x \delta^2 + b_x \delta + 4a_y \delta^2 + 2b_y \delta = p'_{L;1,2}$, where $p'_{L;i,j} = p_{L;i,j} - (b_x + b_y)$. Solving this system yields the four polynomial constants a_x, b_x, a_y, b_y, from which the constant $c_L = 2(a_x + a_y)$ can be determined. Constant $c = p_{L;0,0}$ is chosen equal to the average of the boundary conditions specified near the origin $(0,0)$; i.e., $c = (p_{L;0,1} + p_{L;1,0})/2$. Thanks to this choice, a conductivity that is homogeneous on the boundary (i.e., $p_{L;0,1} = p_{L;2,1} = p_{L;1,2} = \alpha_B$) results in a conductivity that is homogeneous in the model domain (i.e., $a_x = a_y = 0$, $b_x = b_y = 0$, $c = \alpha_B$).

Because of the extremely coarse discretization, the above-presented approximation is very poor. However, we can systematically derive better approximations. For that purpose, we partition the model domain in four grid blocks, which means that we can honor eight boundary conditions. For this grid-refined domain, we solve Poisson equation $\nabla^2 p_{L-1} = p_L + c_{L-1}$ with the eight boundary conditions for p_{L-1}. To be sure, we do not use the polynomial constants calculated for the one-grid block domain; they will be recalculated. Since function p_L is a polynomial of degree 2, function p_{L-1} is a polynomial of degree 4: $p_{L-1} = (a'_x x^4 + b'_x x^3 + c'_x x^2 + d'_x x)$ plus a similar polynomial in y plus a constant, e. Honoring the eight boundary conditions for A_{L-1} results in a system of eight linear algebraic equations that can be solved to find the eight polynomial constants $a'_x, b'_x, c'_x, d'_x, a'_x, b'_x, c'_x, d'_x$. Function p_L is then $p_L = \nabla^2 p_{L-1} = 12a'_x x^2 + 6b'_x x + c'_x$ plus a similar polynomial in y, while constant e is equal to $e = p_{L-1;\,0,\,0} = (p_{L-1;\,0,\,1} + p_{L-1;\,1,\,0})/2$. Again, thanks to this choice, we find a conductivity field that is homogeneous in the domain if the conductivities on the boundary are homogeneous (i.e., all polynomial constants are equal to zero, except for $e = \alpha_B$).

Continuing this procedure, we introduce polynomials of degree $2L$ (for which we recalculate the polynomial constants obtained for the previous 4th degree polynomial example). For that number of L grid blocks, we consider the sequence of Poisson equations $\nabla^2 p_{L-\ell} = p_{L-\ell+1} + c_{L-\ell}$ for $\ell = 2, 3, \ldots, L-3, L-1$.

For $\ell = L - 1$, we have to solve Poisson equation $\nabla^2 p_1 = p_2 + c_1$, which means that polynomial p_1 has degree $2L$. The $4L$ polynomial constants can be determined from the $4L$ boundary conditions for p_1 specified in the $4L$ boundary points, while the additional constant $p_{1;\,0,\,0} = (p_{1;\,0,\,1} + p_{1;\,1,\,0})/2$ is chosen equal to the average of the two boundary values specified near the origin.

2.4.5   A Double Constraint Methodology

Now, we equate sqrt-conductivity α to the above-determined polynomial p_1; i.e., generalizing to three dimensions we set $\alpha(x,y,z) = p_1(x,y,z)$. From the above-presented polynomial construction, it follows that for given $\alpha_B = \alpha(x_B, y_B, z_B)$ on the boundary with coordinates x_B, y_B, z_B sqrt-conductivity $\alpha(x,y,z)$ is known in the whole model domain. For imposed head boundary condition $h_B = h(x_B, y_B, z_B)$, head $h(x,y,z)$ can then be determined in the whole model domain from groundwater flow equation $\nabla \cdot \alpha^2 \nabla h = 0$.

However, instead of specifying boundary values of the sqrt-conductivity, we may impose normal boundary fluxes $q_B = q_n(x_B, y_B, z_B)$. Doing so, groundwater flow equation $\nabla^2 h \alpha - h \nabla^2 \alpha = 0$, or its mathematical equivalent $\nabla \cdot \alpha^2 \nabla h = 0$ (Eq. 2.4), with both this flux boundary condition and this head boundary condition, results in a partial differential problem from which the boundary sqrt-conductivity $\alpha_B = \alpha(x_B, y_B, z_B)$ as well as the sqrt-conductivities, heads and fluxes in the model domain can be determined. More specifically, assume that boundary condition h_B for head h has been determined from measurements. As a consequence, the normal head gradients $\partial h(\alpha_B)/\partial n$ on the boundary can be determined as a function of function $\alpha_B(x_B, y_B, z_B)$, where argument α in $h(\alpha)$ denotes the dependency of head h on sqrt-conductivity field α (a "function of a function" is a functional; see Sect. 3.4). Also, the normal head gradient $[\partial h(\alpha_B)/\partial n]_B$ on the boundary can be determined as a function of function $\alpha(x,y,z)$. To match this normal head gradient to the measured normal boundary flux $q_B(\alpha_B)$, we apply Darcy's law (Eq. 2.2) to obtain the $q_B(\alpha_B) + \alpha_B^2 [\partial h(\alpha_B)/\partial n]_B = 0$. In other words, we have to find the function $\alpha_B = \alpha(x_B, y_B, z_B)$ that makes the functional $q_B(\alpha_B) + \alpha_B^2 [\partial h(\alpha_B)/\partial n]_B$ equal to zero, thus honoring Darcy's law. Thanks to the polynomial construction presented in Sect. 2.4.4, this formulation results in a system of nonlinear algebraic equations for the N_B values of α_B in the N_B boundary points, where $N_B = 2nL$ for n-dimensional flow with $n \geq 2$; see Sect. 2.4.4. Finally, sqrt-conductivity $\alpha(x, y, z)$, head $h(x, y, z)$, and flux density $\mathbf{q}(x, y, z)$ follow from the polynomial expansions.

As an introduction to the double constraint method presented in Chap. 3, we extend the above-presented argument to a constructive iterative approach to find a solution. For an estimated initial sqrt-conductivity $\alpha_{B;i} = \alpha_i(x_B, y_B, z_B)$, the sqrt-conductivity $\alpha_i(x, y, z)$ can be determined from the polynomial construction. Head boundary condition $h_B = h(x_B, y_B, z_B)$ is determined by measurements, and given the esti-mated initial sqrt-conductivity $\alpha_i(x, y, z)$, head $h_i(x, y, z)$ can be calculated in the whole model domain from groundwater flow equation $\nabla \cdot \alpha_i^2 \nabla h_i = 0$ (Eq. 2.4). As a consequence, the normal head derivative $(\partial h_i/\partial n)_B$ on the boundary can be deter-mined, while the normal boundary flux follows from Darcy's law (Eq. 2.2) yielding $q_{B;i} = -\alpha_{B;i}^2 (\partial h_i/\partial n)_B$. The thus-determined normal boundary flux does not gener-ally match the measured normal boundary flux q_B. To determine a new normal boundary flux $q_{B;\,i+1}$ that better matches the measured normal flux, we introduce a new sqrt-conductivity, α_{i+1} by applying Darcy's law in the form $q_B = -\alpha_{B;\,i+1}^2$ $(\partial h_i/\partial n)_B$. Substitution of the above-presented expression for $q_{B;i}$ leads then to the update rule $k_{B;\,i+1} = k_{B;\,i} \, q_B/q_{B;\,i}$ for the conductivity on the boundary. The updated $\alpha_{i+1} = \pm k_{i+1}^{1/2}$ in the whole model domain follows then from the above-derived polynomial construction. After starting the procedure for $i = 0$, we repeat this pro-cedure for $i = 1, 2, 3, \ldots$ until α_{i+1} has converged sufficiently close to α_i.

Until now, we have supposed that the system is well posed (existence of stable unique solution). However, in practical applications, the step $\alpha_{i+1} = \pm k_{i+1}^{1/2}$, which is required for the polynomial construction, may turn out to fail. During the iteration process, k_{i+1} may become negative in a number of boundary points, which means that, even if α_i was a real number, α_{i+1} is no longer real (is an imaginary number). Negative conductivity updates may be caused by measurement errors and also by a poor initial guess of the initial conductivities. To handle such cases, the update rule has to be modified to $k_{B;i+1} = k_{B;i} |q_B/q_{B;i}|$, as will be explained in Chap. 3, Sect. 3.4. Using this modified update rule, negative conductivities caused by a poor initial guess may gradually disappear during the iteration process, but negative conductivities caused by measurement errors cannot be removed.

The above-presented iterative approach may be considered as a simple example of the double constraint methodology presented in Chap. 3. The introduction of smoothness (differentiable polynomials) has resulted in a simplified double con-straint methodology in which the update rule is only applied to the boundary conductivities. Moreover, in this case, the groundwater flow equation is only used to determine the head, while in the general approach this equation has also to be solved for the flux. For a detailed explanation, see Chap. 3.

2.4.6 The One-Dimensional Case

In one dimension, for example, flow through a tube of length L_{tube} filled with porous material, the situation is different. Since there are only two boundary points $(N_B = 2)$, only two polynomial coefficients can be determined, which means that the smoothest possible polynomial is $\alpha = bx + c$. From groundwater flow equation $h\nabla^2\alpha = \nabla^2 h\alpha$, it follows that $d^2 h\alpha/dx^2 = 0$, i.e., $h\alpha = b'x + c'$, $h = (b'x + c')/(bx + c)$, $dh/dx = [b'(bx + c) - b(b'x + c')]/(bx + c)^2$, and $q = -\alpha^2 dh/dx = b(b'x + c') - b'(bx + c)$. Continuity equation $dq/dx = 0$ requires that q is constant, which means that $b' = 0$ or $b = 0$. Solution $b' = 0$ means that there is no head gradient and, as a consequence, no flow, which contradicts our requirement that there should be at least one point in which the flux is nonzero. Hence, the solution is $b = 0$, which means that conductivity $\alpha^2 = k$ is constant. Darcy's law then results in $k = qL_{tube}/(h_{+L_{tube}/2} - h_{-L_{tube}/2})$.

One of the aspects of the Calderón problem is to determine the smoothness requirements in such a way that there exists a unique conductivity field determined by the head–flux boundary conditions (another aspect is to find solutions). Requiring, in addition, that the smoothness condition is as simple as possible, the above-presented example shows that the required smoothness for a one-dimensional Calderón problem is a constant conductivity.

In Sects. 2.4.3, 2.4.4, and 2.4.5, we have shown that for n-dimensional problems with $n \geq 2$ a solution exists. The above-presented "discrete analytical" construction may be considered as an argument for the validity of Calderón's conjecture. In our example, where the N_B boundary points are evenly distributed over the boundary, the continuum limit $N_B \to \infty$ could be considered as a strong argument for the validity of Calderón's conjecture. However, it is not a general proof for all types of bounded continuous space; the smoothness conditions, the distribution of boundary points, and the way of approaching the continuum limit may be specified in different ways. For instance, Kenig et al. (2007) present a three-dimensional problem in which on some parts of the boundary no head–flux data are specified. For more details on the general continuous problem, see, for instance, Barber and Brown (1984, 1986), Sylvester and Uhlmann (1987), Nachman (1996), Brown (1996), Brown and Uhlmann (1997), Siltanen et al. (2000), Bukhgeim and Uhlmann (2002), Uhlmann (2003, 2009), Borcea (2002, 2003), Borcea et al. (2003), Bukhgeim and Uhlmann (2002).

References

Barber D, Brown B (1984) Applied potential tomography. J Phys E: Sci Instrum 17:723–733

Barber DC, Brown BH (1986) Recent developments in applied potential tomography-APT. In: Bacharach SL (ed) Information processing in medical imaging. Martinus Nijhoff, Amsterdam, pp 106–121

Bear J (1972) Dynamics of fluids in porous materials. American Elsevier Publ. Co., New York

Borcea L (2002) Electrical impedance tomography. Inverse Prob 18:99–136

Borcea L (2003) Addendum to electrical impedance tomography. Inverse Prob 19:9978

Borcea L, Gray GA, Zhang Y (2003) Variationally constrained numerical solution of electrical impedance tomography. Inverse Prob 19:1159–1184. PII: S0266-5611(03)57791-7

Bresciani E, Gleeson T, Goderniaux P, de Dreuzy JR, Werner AD, Wörman A, Zijl W, Batelaan O (2016) Groundwater flow systems theory: research challenges beyond the specified-head top boundary condition. Hydrogeol J 24:1087–1090. https://doi.org/10.1007/s10040-016-1397-8

Brown RM (1996) Global uniqueness in the impedance imaging problem for less regular conductivities. SIAM J Math Anal 27:1049–1056

Brown RM, Uhlmann G (1997) Uniqueness in the inverse conductivity problem for non-smooth conductivities in two dimensions. Commun Part Diff Eqns 22:1009–1027

Bukhgeim AL, Uhlmann G (2002) Recovering a potential from partial Cauchy data. Comm Partial Diff Eqns 27:653–668

Butkov E (1973) Mathematical physics. Addison Wesley Publ. Comp., Reading

Calderón AP (1980) On an inverse boundary value problem. Seminar on numerical analysis and its applications to continuum physics (Rio de Janeiro, 1980). Soc Brasil Mat, 65–73, Rio de Janeiro. Also see the reprint: Calderón AP (2006) On an inverse boundary value problem. Comput Appl Math 25(2–3):133–138

Chavent G, Jaffré J (1986) Mathematical models and finite elements for reservoir simulation. Elsevier, North-Holland, Amsterdam

Chen Z, Zhang Y (2009) Well flow models for various numerical methods. Int J Num Anal Mod 6 (3):375–388

Cheney M, Isaacson D, Newell JC (1999) Electrical impedance tomography. SIAM Rev 41:85–101

Duvaut G, Lions JL (1976) Inequalities in mechanics and physics. Springer, Berlin. ISBN 978-3-642-66167-9 (Print) 978-3-642-66165-5 (Online)

El-Rawy M, Batelaan O, Zijl W (2015) Simple hydraulic conductivity estimation by the Kalman filtered double constraint method. Groundwater 53(3):401–413. https://doi.org/10.1111/gwat. 12217

El-Rawy M, De Smedt F, Batelaan O, Schneidewind U, Huysmans M, Zijl W (2016) Hydrodynamics of porous formations: Simple indices for calibration and identification of spatio-temporal scales. Mar Petrol Geol 78:690–700. https://doi.org/10.1016/j.marpetgeo.2016. 08.018

De Smedt F, Zijl W (2018) Two- and three-dimensional flow of groundwater. CRC Press, Taylor & Francis Group, Boca Raton

Holder DS (2005) Electrical impedance tomography. IOP Publishing, Bristol

Kenig C, Sjöstrand J, Uhlmann G (2007) The Calderón problem with partial data. Ann Math 165:567–591

Morse PM, Feshbach H (1953) Methods of theoretical physics. McGraw-Hill, New York

Nachman AI (1996) Global uniqueness for a two-dimensional inverse boundary problem. Ann Math 143:71–96

Olmstead WE (1968) Force relationships and integral representations for the viscous hydrodynamical equations. Arch Ration Mech Anal 31:380–390

Peaceman DW (1977a) Interpretation of well-block pressures in numerical reservoir simulation. SPE 6893, 52nd annual fall technical conference and exhibition, Denver

Peaceman DW (June 1983) Interpretation of well-block pressures in numerical reservoir simulation with non-square grid blocks and anisotropic permeability. Soc Pet Eng J: 531–543

Peaceman DW (Feb 1991) Presentation of a horizontal well in numerical reservoir simulation. SPE 21217, presented at 11th SPE symposium on reservoir simulation in Ananheim, California, 17–20

Salo M (2008) Calderón problem. Lecture notes, Spring 2008 Mikko Salo, Department of Mathematics and Statistics, University of Helsinki. http://users.jyu.fi/~ salomi/lecturenotes/ calderon_lectures.pdf

Siltanen S, Mueller JL, Isaacson D (2000) An implementation of the reconstruction algorithm of A. Nachman for the 2-D inverse conductivity problem. Inverse Prob 16:681–699

Stefanescu SS (1950) Theoretical models of heterogeneous media for electrical prospecting methods with direct currents (in French). Comitetul Geologic, Studii Technice si Economice, Seria D, Nr. 2, Studii Technice si Economice, Imprimeria National, Bucuresti: 51–71

Sylvester J, Uhlmann G (1987) A global uniqueness theorem for an inverse boundary value problem. Ann Math 125:153–169

Tóth J (2009) Gravitational systems of groundwater flow. Cambridge University Press, Cambridge

Trykozko A, Zijl W, Bossavit A (2001) Nodal and mixed finite elements for the numerical homogenization of 3D permeability. Comput Geosci 5:61–64

Uhlmann G (2003) Electrical impedance tomography and Calderon's problem. Med Eng Phys 25:79–90 https://www.math.washington.edu/ ~ gunther/publications/Papers/calderoniprevised.pdf

Uhlmann G (2009) Electrical impedance tomography and Calderón's problem

Inverse Problems 25:123011 (39 p) doi:https://doi.org/10.1088/0266-5611/25/12/123011

http://www.dim.uchile.cl/ ~ axosses/calderoniprevised.pdf

Yeh WWG (1986) Review of parameter identification procedures in groundwater hydrology: The inverse problem. Water Resour Res 22(2):95–108

Zijl W, Nawalany M (1993) Natural groundwater flow. Lewis Publishers, Boca Raton

Chapter 3
The Pointwise Double Constraint Methodology

3.1 Parameter Estimation

The majority of inverse groundwater flow models are based on a "flux model"; that is, a model based on groundwater flow equation (Eq. 2.4) with imposed fluxes on the top boundary (the water table) and on as many as possible other boundaries where fluxes are known, in particular the zero flux at the impervious bottom boundary. The nonzero recharge fluxes on the water table are measured indirectly by using models that take direct measurements of rainfall, runoff, and evapotranspiration into account. Other nonzero fluxes may come from flow rate measurements in production or injection wells. The conductivity field is then determined by manipulating the conductivities in a number of spatial points (grid volumes in a discretized model), in such a way that the heads measured in the observation wells match with the heads calculated by the model. This manipulation is often accomplished automatically by minimization of an objective function (cost function). This minimization results in a gradient matrix or sensitivity matrix. Handling the gradient matrix is a relatively heavy burden, both from a computational and from a programming point of view. Therefore, there is a growing interest in gradient-free calibration methods; see, for instance, Chen and Zhang (2006), Naevdal et al. (2002), who applied the ensemble Kalman filter (EnKF) for that purpose.

One of the approaches to gradient-free calibration is the double constraint methodology introduced in this chapter. This method determines conductivities by using both a model with imposed fluxes (the conventional flux model) and a model with imposed heads (the head model).

© The Author(s) 2018
W. Zijl et al., *The Double Constraint Inversion Methodology*,
SpringerBriefs in Applied Sciences and Technology,
https://doi.org/10.1007/978-3-319-71342-7_3

3.2 Linear Conductivity Estimation

Conventional approaches are based on a flux model and an inversion model based on the flux model's gradient matrix. In contrast, we base our method on the groundwater flow equation (Eq. 2.4) with all known ("measured") *fluxes* imposed as boundary condition (the flux model), and, again, on Eq. 2.4, but now with measured *heads* imposed as boundary condition (the head model). This means that in the head model, the heads measured in the observation wells are imposed. Wexler et al. (1985) originally introduced such an approach for electrical impedance tomography for geophysical and medical imaging. The two solutions of Eq. 2.4, one of the flux model and one of the head model, constrain the inverse modeling process, which is aptly expressed by the name "double constraint method" coined by Yorkey and Webster (1987), Yorkey et al. (1987), Webster (1990).

The double constraint method, which is conceptually very simple and can easily be implemented, is based on the following four steps:

1. In each "point" (center of a grid volume) of the model domain, the three principal conductivities $k_{x;i}$, $k_{y;i}$, and $k_{z;i}$ are assumed to be known initially (index $i = 0$). The initial conductivities may come from "hydrogeological perceptions," i.e., from the knowledge that the hydrogeologist already has about the subsurface under consideration.
2. One constraining forward run—the so-called flux run—is based on the known flux boundary conditions (e.g., recharge rates and production or injection rates of wells) and results in the flux densities $q^F_{x;i}$, $q^F_{y;i}$, and $q^F_{z;i}$ in each point (superscript F denotes results from the flux run).
3. A second constraining forward run—the so-called head run—is based on the known head boundary conditions (including the heads measured in the monitoring wells!) and results in the heads h^H_i as well in as the head gradients $e^H_{x;i} = -\partial h^H/\partial x$, $e^H_{y;i} = -\partial h^H/\partial y$, and $e^H_{z;i} = -\partial h^H/\partial z$ in each point (superscript H denotes results from the flux run).
4. Now the "old" conductivities $k_{x;i}$, $k_{y;i}$, and $k_{z;i}$ are "forgotten," and we determine the updated conductivities $k_{x;i+1}$, $k_{y;i+1}$, and $k_{z;i+1}$ from Darcy's law $k_{x;i+1} = q^F_{x;i}/e^H_{x;i}$, $k_{y;i+1} = q^F_{y;i}/e^H_{y;i}$, and $k_{z;i+1} = q^F_{z;i}/e^H_{z;i}$. Substitution of $e^H_{x;i} = q^H_{x;i}/k_{x;i}$, $e^H_{y;i} = q^H_{y;i}/k_{y;i}$, and $e^H_{z;i} = q^H_{z;i}/k_{z;i}$ results in the following update rule in each point (grid volume center) of the model domain

$$k_{x;i+1} = k_{x;i}\frac{q^F_{x;i}}{q^H_{x;i}}, \ k_{y;i+1} = k_{y;i}\frac{q^F_{y;i}}{q^H_{y;i}}, \ k_{z;i+1} = k_{z;i}\frac{q^F_{z;i}}{q^H_{z;i}} \qquad (3.1)$$

These conductivities, derived as a result of the DCM approach and used in a groundwater flow model, comply with both the flux and the head boundary conditions including the heads measured in the monitoring wells.

The points where $|q_{x;i}^F/q_{x;i}^H - 1| \ll 1$, $|q_{y;i}^F/q_{y;i}^H - 1| \ll 1$, and $|q_{z;i}^F/q_{z;i}^H - 1| \ll 1$ are situated in the so-called terra incognita, i.e., in regions of the model domain which are too far away from locations where the imposed flux and head data can determine the conductivities by update rule Eq. 3.1. On the other hand, the points where $|q_{x;i}^F/q_{x;i}^H - 1| \gg 0$, $|q_{y;i}^F/q_{y;i}^H - 1| \gg 0$, and $|q_{z;i}^F/q_{z;i}^H - 1| \gg 0$ are situated in the "measurement ranges," i.e., in regions sufficiently close to locations where the imposed flux and head data can influence the hydraulic conductivity via the update rule Eq. 3.1.

If the measurement errors of the imposed heads and fluxes would be sufficiently small, and if the initial hydraulic conductivities $k_{x;i}$, $k_{y;i}$, and $k_{z;i}$ would have values that are sufficiently close to the true values, the update equations (Eq. 3.1) would result in nonnegative values of the updated hydraulic conductivities $k_{x;i+1}$, $k_{y;i+1}$, and $k_{z;i+1}$. Under such ideal conditions, a solution of the inverse problem is found in one stroke, without subsequent iterations. However, in practical hydrogeological problems, the quality of these data is generally insufficient to yield a linear inverse problem (one-dimensional flow is a trivial exception). The imposed heads and fluxes are far from error-free, and even if the hydrogeological setting is known beforehand (e.g., the positions of aquifers and aquitards with their characteristic conductivity values), these "soft data" are generally far from sufficiently accurate.

To investigate this aspect, Trykozko et al. (2008) presented a DCM application for estimating a two-dimensional synthetic checkerboard conductivity patterns. Starting with nonnegative conductivities k_0, they noted the occurrence of negative conductivities k_1 after the first DCM stroke. To apply these values for the first iteration step, they used the absolute values $|k_1|$ and it appeared that after the first iteration, the number of negative conductivities k_2 was diminished and so on for further iterations. Iterative improvement of the conductivity pattern diminished the number of negative conductivity values. Although the iterations converged, negative conductivity values did not completely disappear. Also Brouwer et al. (2008), Trykozko et al. (2009), and El-Rawy (2013) have used absolute values of the conductivities. However, they did so without presenting arguments. This book presents an extensive justification of the "absolute value approach"; see Sect. 3.4.

3.3 Generalized Points: The Voxel Notation

Until now, we have based our analysis on the powerful tools of classical mathematical analysis: differentiation, integration, and partial differential equations. In this type of analysis, we consider the differential df of a function f as a very small change in that function, without actually specifying how small that change is. In applied mathematics—the mathematics used by geoscientists and engineers—a differential is generally presented as an "infinitely small" limit to zero: Denoting an arbitrary, not necessarily small difference in function f as Δf, differential df is equal to the limit $\Delta f \to 0$, which means that df is nonzero but approaches zero without specifying how close it is to zero. For reasons that will be explained in Sect. 3.4, we apply this interpretation of differentiation to replace our analytical equations with algebraic equations for our groundwater flow problems.

More specifically, we quantize (discretize) the model space by a grid with a large number of N_V grid volumes with (extremely small but finite) size ΔV. To each grid volume, we assign a volume number, n_V, relating the volume to the coordinates (x, y, z) in its center ($n_V = 1, 2, \ldots, N_V$). Doing so a function $f(x, y, z)$ in point (x, y, z) will be denoted as f_{n_V}. This notation is now extended to vector functions $\mathbf{u} = (u_x, u_y, u_z)$ in point (x, y, z) by the introduction of "generalized grid volumes," or voxels (the term voxel is a portmanteau for volume and pixel, where pixel is a combination of picture and element). A voxel represents just one quantity; for instance, u_x in a grid volume is a voxel while u_y in the same volume is another voxel and u_z in the same volume is again another voxel. In other words, we quantize (discretize) the model space as a uniformly arranged grid with a large number of $N = 3N_V$ voxels. To each grid volume, we assign a voxel number, n, in such a way that vector component u_z has voxel number $n = 3n_V$, component u_y has voxel number $n = 3n_V - 1$, while component u_x has voxel number $n = 3n_V - 2$ ($n = 1, 2, \ldots, N$). Doing so vector function $\mathbf{u}(x, y, z)$ in point (x, y, z) will be denoted as u_n. The thus-introduced notation is useful for discretized models in which the partial differential equations with boundary conditions are replaced with algebraic approximations; also see Narasimhan (2010) as well as Shinbrod's introduction of the quantized Navier–Stokes equations (Shinbrod 1973, part II).

As a result, instead of the N_V volumes with three conductivity components, we consider $N = 3N_V$ voxels with voxel number $n = 1, 2, \ldots, N$. Update rule Eq. 3.1 can now be written as only one equation.

$$ k_{n;i+1} = k_{k_{n;i}} \frac{q_{n;i}^F}{q_{n;i}^H} \tag{3.2} $$

As has already been described in Sect. 3.1, Eqs. 3.1 and 3.2 may result in negative conductivities $k_{n;i+1}$ in a number of voxels. Intuitively, it makes sense to modify the update equation (Eq. 3.2) by replacing $q_{n;i}^F/q_{n;i}^H$ with its absolute value $|q_{n;i}^F/q_{n;i}^H|$, as has been proposed and applied by Brouwer et al. (2008), Trykozko et al. (2008, 2009), and El-Rawy (2013). In Sect. 3.4, we present an elaborate justification of this approach.

Equations 3.1 and 3.2 can be implemented relatively simply, without much additional programming. The flux model and the head model can be run with standard modeling software, without need for software modifications; in our case studies (Chaps. 5 and 7), we have applied MODFLOW. For each grid volume, the (absolute values of the) fluxes obtained as output from the flux model have to be divided by the (absolute values of the) fluxes obtained as output from the head model, after which the model's conductivities have to be updated using Eqs. 3.1 and 3.2. This can simply be done by MATLAB (2017) or another simple computer program.

3.4   Minimization Based on Sqrt-Conductivities

To automate the search for as much as possible nonnegative values of $k_{n;i+1}$, we derive a slightly modified update rule. Instead of basing our considerations directly on the conductivities k_n, we base ourselves on the sqrt-conductivities $\alpha_n = \pm k_n^{1/2}$. That is, we write Darcy's law (2.2) as $\alpha_n^{-1} q_n - \alpha_n e_n = 0$, in which $q_n = q_n^F$ honors the imposed flux conditions while $e_n = e_n^H$ honors the imposed head conditions (note the minus sign in the definitions $e_x = -\partial h/\partial x$, $e_y = -\partial h/\partial y$, and $e_z = -\partial h/\partial z$). Doing so, we accept conditions in which Darcy's law cannot be honored exactly everywhere in the model domain (because $k_n = \alpha_n^2$ cannot become negative), i.e., $\alpha_n^{-1} q_n - \alpha_n e_n \neq 0$.

To keep the error in Darcy's law sufficiently small, we minimize the following function of the N sqrt-conductivities $\alpha_1, \ldots, \alpha_N$ (Kohn and Vogelius 1987; Kohn and McKenney 1990).

$$R(\alpha_1, \ldots, \alpha_N) = \sum_{n=1}^{N} \left(\frac{q_n(\alpha_1, \ldots, \alpha_N)}{\alpha_n} - \alpha_n e_n(\alpha_1, \ldots, \alpha_N) \right)^2 \qquad (3.3)$$

If Darcy's law is honored exactly, the "Darcy residual" $R(\alpha_1, \ldots, \alpha_N)$ is equal to zero.

In the continuum limit $\Delta V_n \to 0$, $N \to \infty$, Eq. 3.3 can be written as

$$R(\boldsymbol{\alpha}) = \iiint\limits_{D} \left(\frac{\mathbf{q}(\alpha, \mathbf{x})}{\alpha(\mathbf{x})} - \alpha(\mathbf{x}) \, \mathbf{e}(\alpha, \mathbf{x}) \right)^2 w(\mathbf{x}) \, \mathrm{d}x\mathrm{d}y\mathrm{d}z \qquad (3.4)$$

where $\mathbf{x}$ is the vector with coordinates x, y, z; D denotes integration over the model domain and $w(\mathbf{x})$ is a weighting function; in the discrete approximation of the above integral $w(\mathbf{x}_n) = w_n = 1/\Delta V_n$ yields Eq. 3.3. To be more precise, Eq. 3.4 represents the continuum formulation for an isotropic porous medium, but extension to anisotropy is straightforward. Darcy residual $R(\boldsymbol{\alpha})$ may be considered as a function of array $\boldsymbol{\alpha}$ containing the sqrt-conductivities in all points of the model domain (an infinitely large number); these sqrt-conductivities have to be determined in such a way that $R(\boldsymbol{\alpha})$ is minimized. The integrand $\ell(\alpha; \mathbf{x}) = (\alpha^{-1}\mathbf{q} - \alpha\mathbf{e})^2$ is a functional, i.e., a "function" of function $\alpha(\mathbf{x})$ (Morse and Feshbach 1953: 275–280; Butkov 1973: 553–562). For practical problems, and in particular for minimization problems, derivations based on an algebraic formulation with a finite number of sqrt-conductivities, as presented by Eq. 3.3, are simpler than derivations based on functional analysis. They are also more insightful because they relate directly to numerical models—in which the model domain is discretized—which play a dominant role in present-day applied geoscience and engineering.

At the stationary points (minimums, maximums, or saddle points) of R, we have $\partial R/\partial\alpha_n = 0$, which results in the following N equations for the N unknowns $\alpha_1, \alpha_2, \ldots, \alpha_n, \ldots, \alpha_{N-1}, \alpha_N$

$$\frac{1}{2}\frac{\alpha_n\partial R}{\partial\alpha_n} = \frac{e_n^2}{\alpha_n^2}\left(\alpha_n^2 - \frac{q_n}{e_n}\right)\left(\alpha_n^2 + \frac{q_n}{e_n}\right)$$

$$+ \sum_{m=1}^{N} e_m\left(\alpha_m^2 - \frac{q_m}{e_m}\right)\left(\frac{\alpha_n\partial e_m}{\partial\alpha_n} - \frac{1}{\alpha_m^2}\frac{\alpha_n\partial q_m}{\partial\alpha_n}\right) = 0 \tag{3.5}$$

In voxels where $q_n/e_n > 0$ (flow obtained by flux model has the same direction as flow obtained from head model), we can honor Darcy's law $\alpha_n^2 = q_n/e_n > 0$ exactly. If this is the case in all voxels, substitution into Eq. 3.3 shows that $R = 0$. However, if there are voxels where $q_n/e_n < 0$ (flow obtained by flux model has direction opposite to flow obtained from head model), which is generally the case for a minority of voxels (Trykozko et al. 2008, 2009), a solution has to be derived from the above system of nonlinear algebraic equations.

In general, such systems have many solutions. Only one of the stationary points represents the global minimum of $R(\alpha_1, \ldots, \alpha_N)$, while the others are either a local minimum, or a maximum, or a saddle point. The gradient matrices $q'_{nm} = \alpha_n\partial q_m/\partial\alpha_n$ [L T^{-1}] and $e'_{nm} = \alpha_n\partial e_m/\partial\alpha_n$ [−] reflect the sensitivity of voxel conductivity α_n to variations in other voxel conductivities.

We solve Eq. 3.5 approximately by neglecting gradient matrices q'_{nm} and e'_{nm}, which results in

$$\left(\alpha_n^2 - \frac{q_n}{e_n}\right)\left(\alpha_n^2 + \frac{q_n}{e_n}\right) = 0 \tag{3.6}$$

To correct for the neglected gradient matrices, Eq. 3.6 is applied iteratively. Starting with initial conductivities $k_{n;i} = \alpha_{n;i}^2$, the numerical model calculates $q_{n;i}$ (obtained from the flux run) and $e_{n;i}$ (obtained from the head run). The solution of system (3.6) is

$$k_{n;i+1} = \frac{|q_{n;i}|}{|e_{n;i}|} = k_{k_{n;i}}\frac{|q_{n;i}^F|}{|q_{n;i}^H|} \tag{3.7}$$

Written for the three directions x, y, z and adding an additional term to handle zero fluxes, Eq. 3.7 can be written for each grid volume n as

$$k_{x;i+1} = k_{x;i}\frac{|q_{x;i}^F|}{|q_{x;i}^H|}, \; k_{y;i+1} = k_{y;i}\frac{|q_{y;i}^F|}{|q_{y;i}^H|}, \; k_{z;i+1} = k_{z;i}\frac{|q_{z;i}^F|}{|q_{z;i}^H|} \tag{3.8}$$

Comparison of Eqs. 3.7 and 3.8 with Eqs. 3.2 and 3.1, respectively, shows that we end up with almost the same update rules; the difference is that not the fluxes and head gradients, but their absolute values have to be used.

In this approximation, the Hessian is a diagonal matrix with positive components $\partial^2 R_{i+1}/\partial \alpha_{i+1}^2 = 8e_{n;i}^2 > 0$, which means that all stationary points of Darcy residual $R_{i+1}\left(\alpha_{1;i+1}, \ldots, \alpha_{N;i+1}\right) = \sum_{n=1}^{N}\left(\alpha_{n;i+1}^{-1} q_{n;i} - \alpha_{n;i+1} e_{n;i}\right)^2$ are minimums.

To avoid division by zero, we may replace Eq. 3.7 with $k_{n;i+1} = k_{n;i}\left(|q_{n;i}^H| + q_n^{\varepsilon}\right)/\left(|q_{n;i}^H| + q_n^{\varepsilon}\right)$, where q_n^{ε} is a small flux with respect to the representative flux in the model domain or a subdomain.

3.5   Arguments for the Neglect of Gradient Matrices

The most important argument for neglecting the terms with the gradient matrices in Eq. 3.5 is that the handling of these matrices is extremely demanding in terms of computer memory storage and computational time. Moreover, to construct these matrices, additional software has to be developed, while neglecting them allows for almost no additional programming when a standard groundwater model like, for instance, MODFLOW is applied.

Below we will present a second argument why the global minimum may result in a conductivity field that is too smooth to be realistic. In other words, we present an argument why the inverse solution obtained by approximation Eq. 3.6 may be preferred above a solution obtained by Eq. 3.5. For that purpose, we consider the two gradient matrices $e'_{nm} = \alpha_n \partial e_m / \partial \alpha_n$ and $q'_{nm} = \alpha_n \partial q_m / \partial \alpha_n$ in the second term right-hand side of Eq. 3.5. When all sqrt-conductivities α_n $(n = 1, 2, \ldots, N)$ are multiplied by a constant c, the gradient matrices do not change. This means that a change of the sqrt-conductivity field (an update) caused by a change in flux field (flux run) and head gradient field (head run) changes the matrices only when the ratios between sqrt-conductivities α_n and α_m $(n \neq m)$ are changed. Consequently, these matrices play a role in updating the *ratios* between the different α_n-values.

This ratio-changing role of the matrices is exemplified for one-dimensional flow (e.g., flow through a tube filled with porous material (El-Rawy et al. 2015)). For the case $q/e_n > 0$, update rule Eq. 3.5 results in a Darcy residual equal to zero. Since in one-dimensional flow the flux density $q_n = q$ is the same in each voxel n, the case $q/e_n < 0$ is obviously based on a measurement error (for instance, the sign of the heads at the two end points of the tube has been interchanged). Since the flux depends only on the flux imposed at the boundary (on one of the two end points of the tube) and is independent from the conductivities, matrix q'_{nm} is equal to zero. Let us now consider a homogeneous conductivity, i.e., the case in which all voxel conductivities have the same value. In that case, the head gradients in the voxels are everywhere the same and are independent from the conductivity values. Hence, the head gradient depends only on the heads imposed at the boundaries (the two end points of the tube), which means that matrix e'_{nm} is equal to zero and the Darcy residual has reached its minimum. If we would start with a heterogeneous conductivity, the Darcy residual is larger than its minimum. Only if the gradient matrices are equal to zero, i.e., if the conductivity has been smoothed to a homogeneous field, the Darcy residual has reached its minimum. However, why should we care for the exact estimation of a minimum in case of obvious measurement errors? When neglecting the gradient terms in Eq. 3.5 from the beginning, we end up with the same solution as we would obtain for $q/e_n > 0$, which is, in fact, the correct solution.

In many practical cases, it is desirable to preserve ratios between conductivities in neighboring grid volumes as much as possible. Uncontrolled spatial smoothing of conductivities may lead to unwanted disappearance of conductivity contrasts that have to be preserved, for instance, the conductivity contrasts between well-conducting and poorly conducting subsurface bodies. Some conductivity smoothing may be required in more or less homogeneous parts of the subsurface (homogeneous in an averaged sense, heterogeneous in the fine-scale details) and has to be guided, for instance, by zonation (see Chap. 6). However, we have to avoid unguided smoothing in order to preserve the soft data about differences in rock-type conductivity of different geological formations. Since taking the gradient matrices into account leads to unguided smoothing, neglect of the gradient terms in Eq. 3.5 does not only liberate us from heavy computational requirements, but also liberate us from unwanted smoothing.

As a third argument to omit the gradient matrices can be based on the fact that if the stationary point of Darcy residual Eq. 3.3 or 3.4 is a minimum, it is generally not the global, but a local minimum. The probability that we have found the global minimum is very small. So, why not being satisfied with the approximate local minimum obtained by omitting the gradient matrices?

3.6   Anisotropy

The anisotropy of the updated conductivities k_x, k_y, and k_z (omitting iteration index $i + 1$) may be too extreme. For the grid volumes in which the anisotropy ratios are outside the pre-ordained intervals $a_{xz} \leq k_x/k_z \leq b_{xz}$, $a_{yz} \leq k_y/k_z \leq b_{yz}$, and $a_{xy} \leq k_x/k_y \leq b_{xx}$, we determine new conductivities by transforming the updated conductivities to conductivities $\widetilde{k}_x$, $\widetilde{k}_y$, and $\widetilde{k}_z$ using the following "mixing rule."

$$\begin{pmatrix} \widetilde{k}_x \\ \widetilde{k}_y \\ \widetilde{k}_z \end{pmatrix} = \begin{pmatrix} \eta_{xz} \\ \eta_{yz} \\ 1 \end{pmatrix} \left(\frac{k_x}{\eta_{xz}}\right)^{\beta_x} \left(\frac{k_y}{\eta_{yz}}\right)^{\beta_y} (k_z)^{\beta_z} \qquad . \quad (3.9)$$

where $\eta_{xz} = (a_{xz}b_{xz})^{1/2}$ and $\eta_{yz} = (a_{yz}b_{yz})^{1/2}$ are the geometric mean values of the pre-ordained intervals while β_x, β_y, and β_z are given by

$$\begin{pmatrix} \beta_x \\ \beta_y \\ \beta_z \end{pmatrix} = \frac{1}{|q_x^F| + |q_y^F| + |q_z^F|} \begin{pmatrix} |q_x^F| \\ |q_y^F| \\ |q_z^F| \end{pmatrix} \tag{3.10}$$

For flow almost exclusively in the x direction $q_y^F \approx q_y^H \approx 0$ and $q_z^F \approx q_z^H \approx 0$, which means that Eq. 3.7 or 3.8 cannot accurately update the values of k_y and k_z; only k_x can be updated accurately. In that case, Eq. 3.10 simplifies to $\beta_x \approx 1$, $\beta_y \approx 0$, and $\beta_z \approx 0$. Substitution into Eq. 3.9 yields $\tilde{k}_x = k_x$, $\tilde{k}_y = (\eta_{yz}/\eta_{xz}) k_x$, and $\tilde{k}_z = (1/\eta_{xz}) k_x$, which means that the inaccurate updates k_y and k_z do not play a role and the final result has anisotropy ratios within the pre-ordained bounds. This approach is in agreement with the philosophy upon which streamline methods are based (Nelson 1960, 1961, 1962, 1968; Datta-Gupta et al. 1998).

It is also possible to introduce pre-ordained anisotropy ratios without intervals in which these ratios may vary. For that case, we define sqrt-conductivities $\alpha_x = \eta_{xz}^{1/2}\alpha_z$ and $\alpha_y = \eta_{yz}^{1/2}\alpha_z$ with prescribed η_{xz} and η_{yz}. Substitution into the Darcy residual given by Eq. 3.3 and minimization with respect to α_z take the specified anisotropy ratios directly into account and result in the following update rule.

$$\begin{pmatrix} k_{x;i+1} \\ k_{y;i+1} \\ k_{z;i+1} \end{pmatrix} = \begin{pmatrix} k_{x;i} \\ k_{y;i} \\ k_{z;i} \end{pmatrix} \sqrt{\frac{\frac{(q_{x;i}^F)^2}{\eta_{xz}} + \frac{(q_{y;i}^F)^2}{\eta_{yz}} + (q_{z;i}^F)^2}{\frac{(q_{x;i}^H)^2}{\eta_{xz}} + \frac{(q_{y;i}^H)^2}{\eta_{yz}} + (q_{z;i}^H)^2}} \tag{3.11}$$

For isotropy $(\eta_{xz} = \eta_{yz} = 1)$, update rule Eq. 3.11 simplifies to the result $k_{i+1} = |\mathbf{q}_i|/|\mathbf{e}_i| = k_i|\mathbf{q}_i^F|/|\mathbf{q}_i^H|$. This result was for the first time obtained by Kohn and Vogelius (1987); also see Kohn and McKenney (1990).

Wexler et al. (1985) and Wexler (1988) invented a method in which the residual $\tilde{R}_{i+1}(k_{1;i+1}, \ldots, k_{N;i+1}) = \sum_{n=1}^N (q_{n;i} - k_{n;i+1}e_{n;i})^2$ is minimized with respect to $k_{z;i+1}$, rather than minimizing Darcy residual Eqs. 3.3 and 3.4 with respect to $\alpha_{z;i+1}$. Under the condition that conductivities have to be nonnegative, this minimization results in

$$\begin{pmatrix} k_{x;i+1} \\ k_{y;i+1} \\ k_{z;i+1} \end{pmatrix} = \begin{pmatrix} k_{x;i} \\ k_{y;i} \\ k_{z;i} \end{pmatrix} \frac{q_{x;i}^F q_{x;i}^H + q_{y;i}^F q_{y;i}^H + q_{z;i}^F q_{z;i}^H}{(q_{x;i}^H)^2 + (q_{y;i}^H)^2 + (q_{z;i}^H)^2}$$

$$= \begin{pmatrix} k_{x;i} \\ k_{y;i} \\ k_{z;i} \end{pmatrix} \frac{\mathbf{q}_i^F \cdot \mathbf{q}_i^H}{\mathbf{q}_i^H \cdot \mathbf{q}_i^H} \quad \text{for } \mathbf{q}_i^F \cdot \mathbf{q}_i^H \geq 0 \qquad (3.12)$$

$$\begin{pmatrix} k_{x;i+1} \\ k_{y;i+1} \\ k_{z;i+1} \end{pmatrix} = \begin{pmatrix} 0 \\ 0 \\ 0 \end{pmatrix} \quad \text{for } \mathbf{q}_i^F \cdot \mathbf{q}_i^H < 0$$

The inventors considered electrically conducting isotropic media ($k_x = k_y = k_z = k$) for applications in geophysical and medical imaging including subsurface imaging of pollution plumes in groundwater (Tamburi et al. 1988). In their patent (Fry and Wexler 1995), they describe their invention as a tool for solving the field equation of an electrically conductive medium with the aim to extract an image of its interior based on the electric impedivity distribution in the medium.

Wexler's update rule Eq. 3.12 is attractive because of its simplicity. However, the fact that the minimum of their Darcy residual $\tilde{R}$ may result in conductivities that are equal to zero is a disadvantage, because zero conductivity values may hamper meaningful further iterations. This was avoided by Brouwer et al. (2008), Trykozko et al. (2008, 2009), and El-Rawy (2013) by modifying Eq. 3.12 and replacing the term $\mathbf{q}_i^F \cdot \mathbf{q}_i^H$ with its absolute value $| \mathbf{q}_i^F \cdot \mathbf{q}_i^H |$. El-Rawy (2013) has compared the results obtained by modified update rule Eq. 3.12 with the results obtained by update rule Eqs. 3.9 and 3.10; the differences appear to be small and irrelevant in view of the measurement inaccuracy (especially those of the fluxes) that is generally encountered in hydrogeological problems.

3.7 Convergence and Practical Termination Criterions

The first DCM stroke ($i = 0$) followed by subsequent iterations ($i = 1, 2, 3, \ldots$) can be terminated after a minimum value of R_{i+1} has been reached. In most cases, the thus found minimum will be one of the many local minimums (Chavent 1987). If the number of measurements (imposed flux-head pairs) is much smaller than the number of voxel conductivities, the position of this minimum in the solution space generally depends on the specified initial values of k_i used to calculate q_i and e_i. Instead of finding the minimum of R_{i+1}, which is relatively difficult, we could stop after having found a minimum of the related quantity $\Delta k/k = |(k_{i+1} - k_i)/k_i|^2 = |(q_F/q_H)_{i+1} - 1|^2$ (summed over all grid voxels). In cases where the convergence is monotonous $(k_{i+1} < k_i)$, we stop the iterations after $\Delta k/k < \varepsilon$ in which ε is a small number (e.g., $\varepsilon = 0.01$).

To find the global minimum, we can introduce N_T conductivity types. Each voxel belongs to one of these types. For instance, we can define two types: a well-conducting type with conductivity 0.5 m/day and a poorly conducting type with conductivity 0.005 m/day. For N voxels, there are $(N_T)^N$ different spatial conductivity configurations that can be used as initial condition for the calibration method. The configuration with the smallest minimum is then an approximation of the global minimum. Doing this for thousands to millions of voxels is impossible; it would require an astronomical large number of inverse modeling runs.

However, if we partition the model domain in a limited number of N_L zones, say four zones, we need to calibrate only $(N_T)^{N_L} = 16$ conductivity configurations. In this approach, a zone is a collection of voxels having the same conductivity. The grid volumes belonging to a zone need not necessarily be spatially connected; a zone may consist of spatially disconnected parts. A realistic zonation has to be based on hydrogeological insights obtained from other sources, like geophysical prospecting, sedimentology, pumping well test, and other soft data.

Until now, we have focused on the estimation of the hydraulic conductivities in the grid volumes of numerical models like finite difference and finite element models, the so-called pointwise double constraint method. However, for analytical methods and analytical element methods, the model domain is partitioned in a limited number of zones with homogeneous conductivity. In such cases, the pointwise DCM cannot be used. Moreover, zones with homogeneous conductivity, or with fine-scale heterogeneity superimposed on coarse-scale homogeneity, are often required to avoid too large conductivity contrasts within subsurface bodies with the same rock type. Therefore, Chap. 6 extends the double constraint methodology to estimate homogeneous zone conductivities.

3.8   Boundary Conditions "Upside Down"

Suppose that we consider a problem in which only flux boundary conditions are imposed. Such types of problems are frequently considered in petroleum reservoir simulations, where the fluids (water, oil, gas) are contained in a closed reservoir and flow is caused only by production and injection wells. Provided that the volumetric balance "production rate equals injection rate" (Eq. 2.3) is honored, the flux model results in a well-posed problem. For a specified conductivity field, the fluxes and head gradients can be determined everywhere in the flow field, while the head is known up to a constant. However, some, not all, numerical methods for solving the system of algebraic equations perform better if in one point the head is specified. In the double constraint method, this means that for the flux model, all production/injection rates are imposed except for one well (the "upside-down well") where the well head is imposed. In that case, the head run is based on imposed well heads, except for the "upside-down well," where the volumetric flow rate is imposed. Similar types of problems may be encountered in hydrogeology. Below we analyze the consequences of such a deviation from the original flux model and head model principle.

We simplify the analysis for an isotropic medium for which we minimize a modified version of Darcy residual Eq. 3.3.

$$R(\alpha) = \sum \left(\alpha^{-1}\mathbf{q}^A - \alpha\mathbf{e}^B\right)^2$$
$$= \sum \left[\alpha^{-2}\left(q^A\right)^2 - 2\mathbf{q}^A \cdot \mathbf{e}^B + \alpha^2\left(e^B\right)^2\right] \tag{3.13}$$

where $q^A = |\mathbf{q}^A|$ and $e^B = |\mathbf{e}^B|$, while the superscripts A and B denote types of boundary conditions. In the original DCM, $\mathbf{q}^A$ is equal to flux $\mathbf{q}^F$ obtained from the flux model, while $\mathbf{e}^B$ is head gradient $\mathbf{e}^H = -\nabla h^H$ obtained from the head model. In that case, a flux run based on an initial conductivity field k_0 with a value far away from the true value of k results in an initial flux field $\mathbf{q}_0^F$, while a head run based on that k_0 results in an initial head gradient $\mathbf{e}_0^H$. Minimization of Darcy residual Eq. 3.13 results in $k_1 = q_0^F/e_0^H$. Applied to one-dimensional flow (flow through a tube filled with porous material), this is the final solution with $R(\alpha_1) = 0$; additional iterations do not change this result. For two- and three-dimensional problems, iterations $i = 1, 2, \ldots, I$ result in $k_{i+1} = q_i^F/e_i^H$. The iterations are terminated for sufficiently small $R(\alpha_I)$. Because the ratio q_i^F/e_i^H is bounded by the fluxes and heads imposed on the boundaries, the iterations do not lead to an unbounded value of the final solution k_I.

Now we choose the other extreme. In Eq. 3.13, we choose $\mathbf{q}^A$ as the flux obtained by the head model (i.e., $\mathbf{q}^A = \mathbf{q}^H$), while $\mathbf{e}^B$ is chosen as the head gradient is chosen as the head gradient obtained by the flux model (i.e., $\mathbf{e}^B = \mathbf{e}^F$). Minimization of Darcy residual Eq. 3.13 yields $k_1 = q_0^H/e_0^F$. Expressing this result in the bounded ratio q_0^F/e_0^H yields $k_1/k_0 = k_0e_0^H/q_0^F$. Further iterations result in $k_{i+1}/k_0 = k_0e_0^H/q_0^F \times k_1e_1^H/q_1^F \times \cdots \times k_ie_i^H/q_i^F$ in which the terms e_i^H/q_i^F are bounded by the imposed boundary conditions and have order of magnitude e^H/q^F. This result shows that boundary conditions "upside down" lead to unstable DCM: An initial k_0 smaller than q^F/e^H results in a much smaller k_{i+1}, while an initial k_0 larger than q^F/e^H results in a much larger k_{i+1}. Only if initial k_0 is exactly equal to q^F/e^H (the exact solution), the upside-down DCM does not explode; any small deviation from the exact solution, however small it may be, blows up exponentially. In other words, the upside-down DCM is unstable.

In the above-presented discussion, we have considered two extremes of imposing boundary conditions in Darcy residual Eq. 3.13. The conventional way results in a stable iteration process, while the upside-down approach leads to an exponentially unstable iteration process. The question is: What about a mixed condition under which in residual Eq. 3.13 superscript A is equal to F at some points on the boundary and superscript A is equal to H at some other points of the boundary, while for superscript B the complement of the above-specified conditions holds? Although it is more difficult to analyze this problem, it seems reasonable to assume that, in the best case, such a mixed condition leads to a less accurate solution than the solution obtained by posing the boundary conditions in the conventional way. In the worst case, the mixed solution becomes unstable.

3.9 In Depth: A Closer Look at Imaging and Calibration

3.9.1 Imaging Versus Calibration

Let us, for the sake of argument, consider a hypothetical case study for which we know the real spatial conductivity distribution, as it occurs in the natural subsurface. This real conductivity field is then used as input of a mathematical model based on exactly measured fluxes, as they occur in nature, as boundary conditions. This way we have a perfect flux model.

To be able to find numerical solutions, the mathematical model is based on a discrete approximation of the governing equations (Eqs. 2.1–2.7). For our analysis, we have to make a distinction between flux-continuous approximation methods and head-continuous approximation methods. Flux-continuous methods yield fluxes that are continuous at the grid faces between the grid volumes, while the heads are continuous only at the centers of the grid faces. Models that are based on the block-centered finite difference method are flux continuous. The MODFLOW model (Harbaugh 2005; Harbaugh et al. 2000) is an example and is, therefore, a flux-continuous model. Not only in groundwater hydrology, but also in petroleum reservoir engineering the block-centered finite difference method is the most popular numerical technique (Aziz and Settari 1979; Peaceman 1977). However, at the grid faces, the heads are continuous only at the grid face centers, not at the whole face. On the other hand, head-continuous methods result in heads that are continuous at the grid faces, while the fluxes are discontinuous at the faces. Conventional node-based finite element methods (like, for instance, FEFLOW (Diersch 2005)), are head-continuous methods. In contrast to flux-continuous methods, head-continuous methods result in discontinuous flow velocities at the grid faces, which may result in a less accurate determination of streamlines (Kaasschieter 1990; Kaasschieter and Huijben 1992). For these reasons, so-called streamline methods for estimation of conductivities, generally applied to block-centered finite difference methods, have to take into account the head discontinuities over the faces (Datta-Gupta et al. 1998).

More importantly, flux-continuous models perform the calculation of heads and fluxes in such a way that they underestimate conductivity values, while head-continuous models perform the calculation of heads and fluxes in such a way that they over-estimate conductivity values (except for one-dimensional flow). This under/over-estimation can be proved mathematically, and has been exemplfied for a number of synthetic conductivity patterns (Trykozko et al. 2001; Zijl and Trykozko 2001).

Consequently, although the MODFLOW-based flux model will result in correctly approximated fluxes, the head gradients will be over-estimated. Similarly, a MODFLOW-based head model with the same conductivities results in correctly approximated heads and head gradients; however, the fluxes will be underestimated.

Since the double constraint method (DCM) is based on the correctly approximated fluxes obtained from the flux run divided by the correctly approximated head gradients obtained from the head run, the resulting conductivity will be a good approximation of the initially specified real conductivities. This also means that, even if we have specified initial conductivities that are far removed from the real conductivities, the conductivities obtained after termination of the double constraint method's iterations may be considered as an approximate image of the subsurface conductivity field. In other words, we may consider the double constraint method as a geophysical imaging method, a kind of geo-hydraulic method to determine the hydraulic conductivity pattern of the subsurface. Alternatively, when considering electrical potentials and currents instead of hydraulic heads and fluxes, it is a geo-electrical method to determine the electric conductivity pattern of the subsurface, for instance, to detect interfaces between fresh and saline water. The method is also applicable as inversion technique in electrical impedance tomography for geophysical and medical imaging.

However, many hydrogeological studies do not aim at obtaining a realistic image of the subsurface, but aim at a model—generally a flux model—in which the calculated heads match with the heads measured in the piezometers. In such studies, the head gradients obtained from the flux model with conductivities obtained by the double constraint method will be too large. Therefore, calibration—matching with the measured heads—will result in higher conductivities than obtained by the double constraint method. To be sure, for comprehensive models with thousands to millions of grid blocks, this difference will be small and in the limit of infinitely fine discretization the difference vanishes. For case studies based on comprehensive models, the differences will generally fall within the observation error (see Chap. 4, Sect. 4.2 for more details regarding the observation error). Nevertheless, comparison of the double constraint method with conventional gradient-based calibration methods may show a systematic difference. For instance, if in the flux model an upstream head is imposed, the double constraint method yields conductivities which, when applied to the flux model, will result in lower downstream heads than measured in the downstream piezometers; also see the Schietveld case study in Chap. 6.

For simple models, i.e., coarsely gridded models based on a limited number of grid blocks, the conductivity image obtained by double constraint method may deviate appreciably from the conductivities obtained by calibration. Below we will present a method related to the double constraint method with focus on calibration instead of imaging.

3.9.2 Constrained Back Projection

Below we briefly introduce "constrained back projection," a calibration method based on iterations between two models: the forward model—generally the flux model—and the back projection model. However, unlike the double constraint method, which is an imaging method, constrained back projection is a calibration method because it is only based on a flux model, not on a head model.

3.9.2.1 The Forward Model

First we present the forward model. For flux-continuous discrete approximation methods, the groundwater flow equation (Eq. 2.4) can be written as the matrix-vector equation.

$$S\frac{\partial H}{\partial t} + DZ^{-1}(D^T H + \Pi) = 0 \tag{3.14}$$

while the column vector of fluxes through the faces is given by the discretized (or quantized) analog of Darcy's law (Eq. 2.2).

$$Q = Z^{-1}(D^T H + \Pi) \tag{3.15}$$

Here S [L^2] is a diagonal matrix with components $(s_1 V_1, \ldots, s_n V_n, \ldots, s_N V_N)$ in its diagonal, where s_n [L^{-1}] $(n = 1, 2, \ldots, N)$ represents the specific storages of the N grid volumes with volumes V_n [L^3]; H [L] is a column vector with the heads in the grid volume centers; D [−] is the incidence matrix relating grid volume numbers to grid face numbers; Π [L] is a column vector with heads on the external boundary; Q [$L^3 T^{-1}$] is a column vector of the fluxes through the faces; the flux density q [$L T^{-1}$] through a face is equal to the flux divided by the surface area of that face (in this book, the terms flux and flux density are used as synonyms, except when the difference is specified explicitly); and Z^{-1} [$L^2 T^{-1}$] is the inverse of the resistance matrix (impedance matrix).

The elements of resistance (impedance) matrix Z [$T L^{-2}$] are linear combinations of the grid volume resistivities (or impedivities) $\gamma_{x;n} = k_{x;n}^{-1}$, $\gamma_{y;n} = k_{y;n}^{-1}$, and $\gamma_{z;n} = k_{z;n}^{-1}$ [$T L^{-1}$]. In the block centered finite difference method (MODFLOW), matrix Z is a diagonal matrix, which means that determination of its inverse, Z^{-1}, does not pose computational problems. This method has been worked out in great detail by Mohammed (2009) and Mohammed et al. (2009a), who developed and exemplified this approach for incompressible flow (negligible term $S \partial H / \partial t$). He also compared his approach to Eq. 3.14 with an improved version of the edge-based three-dimensional stream function method initially proposed by Zijl and Nawalany (2004); also see Nawalany and Zijl (2010). For the mixed-hybrid finite element method (or face centered finite element method) with arbitrarily shaped grid volumes and off-diagonal resistivity components, the inverse of the resistance matrix has to be determined volumewise to avoid excessive numerical computations (Chavent and Jaffré 1986; Kaasschieter 1990; Kaasschieter and Huijben 1992; Zijl 2005a, b).

The above-presented "generalized finite differences" approach to discretization has been introduced by Bossavit (1998a, b, 1999, 2000) in the context of electromagnetism; for a good summary, see Bossavit (2005). In the limit of an infinite number of infinitely small discretization volumes, Eqs. 3.14 and 3.15 transform into the well-known continuum equations and boundary conditions presented in Chap. 2 . To be more precise, these equations transform to the equations in the Grassmann-Cartan notation of exterior calculus, in which the metrics of space (the dimensions of the grid) appear only in the resistance matrix. This notation is mathematically equivalent to the more usual Gibbs-Heaviside notation based on the operators gradient ∇, divergence $\nabla\cdot$, and curl $\nabla\times$ in which the metric of space appears in the operators (Frankel 2004; Zijl 2005a, b).

3.9.2.2   The Back Projection Model

Now we turn to the back projection model. Choosing the N initial voxel resistivities $\gamma_{1;i}, \ldots, \gamma_{N;i}$, the initial resistance matrix $Z_i(\gamma_{1;i}, \ldots, \gamma_{N;i})$ can be determined (a voxel is a generalized grid volume; see Chap. 3, Sect. 3.3). The model to be calibrated is a flux model based on Eqs. 3.13 and 3.14, which means that the volume-based heads H_i^F and the face-based fluxes Q_i^F are calculated by the flux run. To update the resistivities, Eq. 3.15 is applied in the form.

$$Z_{i+1} Q_i^F = E_i^{MF} \tag{3.16}$$

$$E_i^{MF} = D^T H_i^{MF} + \Pi_i^{MF} \tag{3.17}$$

where $Z_{i+1}(\gamma_{1;i+1}, \ldots, \gamma_{N;i+1})$ is the resistance matrix composed of the resistivities $\gamma_{1;i+1}, \ldots, \gamma_{N;i+1}$ that have to be updated; Q_i^F is the above-defined vector of fluxes through the faces calculated by the forward flux run. However, H_i^{MF} is the vector of grid block centered heads in which as much as possible measured heads are imposed while the remaining heads are chosen equal to the heads calculated by the flux run. Similarly, vector Π_i^{MF} represents the measured face centered heads on the external boundaries while the remaining heads are equal to the heads calculated by the flux run.

Since the resistance matrix is linear in the grid volume resistivities $\gamma_{1;i+1}, \ldots, \gamma_{N;i+1}$, it makes sense to define resistivity vector $Y_{i+1} = (\gamma_{1;i+1}, \ldots, \gamma_{N;i+1})^T$ and back projection matrix P_i independent from the resistivities, in such a way that

$$P_i Y_{i+1} = Z_{i+1} \left(\gamma_{1;i+1}, \ldots, \gamma_{N;i+1} \right) Q_i^F \tag{3.18}$$

Combination of Eqs. 3.16, 3.17, and 3.18 results in the linear algebraic system.

$$P_i Y_{i+1} = E_i^{MF} \tag{3.19}$$

In Eq. 3.19, zonation can be introduced by setting the resistivities within each zone equal to each other. This way the original system (Eq. 3.19) is transformed into the system.

$$P_i^* Y_{i+1}^* = E_i^{MF} \tag{3.20}$$

Because vector Y_{i+1}^* contains only the zone resistivities, there are less unknown resistivities in system Eq. 3.20 than in system Eq. 3.19 and back projection matrix P_i^* has less columns than matrix P_i. If the number of linearly independent algebraic equations is greater than the number of resulting conductivities matrix $P_i^{*T} P_i^*$ is nonsingular, which means that the updated conductivities follow from solving the least squares system.

$$P_i^{*T} P_i^* Y_{i+1}^* = P_i^{*T} E_i^{MF} \tag{3.21}$$

A system of algebraic equations obtained from a least squares approach is generally ill-conditioned. However, for not too many unknown zone conductivities, a mathematically equivalent system can be solved by orthogonal matrix decomposition (also known as QR decomposition). This method, which has a much better matrix condition, is available in standard numerical mathematical libraries like MATLAB (2017). Negative resistivities can be removed by taking their absolute value, from which we find the positive conductivities $k_{x;n;i+1} = 1/|\gamma_{x;n;i+1}|$, $k_{y;n;i+1} = 1/|\gamma_{y;n;i+1}|$, and $k_{z;n;i+1} = 1/|\gamma_{z;n;i+1}|$.

Van Leeuwen (2005) was inspired by a simpler form of the above-presented approach. She considerably improved the approach and developed and compared different back projection techniques. Thanks to her work a much better choice could be made for a practice-oriented development of the back projection method; also see Zijl (2004, 2007), Mohammed (2009), Mohammed et al. (2009b), Zijl et al. (2010).

In the context of the above-presented approach to calibration, it is appropriate to cite Olsthoorn (1998):

"The direct approach treats the parameters that need to be optimized as the dependent variables and the measured heads and flows as given. It was the only method that was practical before computers became available. Huisman (1950) used a mesh of 52 hexagons to calibrate the aquitard between the phreatic and second aquifer, in the 36-km^2-large Amsterdam Dune Area. He did this by hand. Given net precipitation, the transmissivity, and the head in the phreatic aquifers, he computed the water balance of each hexagon. This flow was then used, together with the head difference between the phreatic and the second aquifer, to compute the hydraulic resistance of the aquitard for each hexagon. By smoothing, he dealt with the high sensitivity of head gradients, used to compute the flows. He succeeded in revealing the general spatial pattern of this resistance."

Comparing Huisman's calibration of vertical hydraulic resistances with resistance calibration by the back projection equation (Eq. 3.21) Huisman's smoothing to deal with high sensitivity of head gradients may be compared with zonation (see Chap. 6) combined with iterations of the head gradient equation (Eq. 3.17).

By historical default, the hydrogeological community, as well as the community of petroleum reservoir engineers, traditionally distrusts direct inversion methods. Their distrust is generally based on the argument that for small head gradients, as they often occur in flat deltaic regions, the solution becomes "unstable"; also see Sect. 2.4.1 and Olsthoorn (1998). Indeed, at locations where there is no flow Q_i^F is equal to zero and, consequently, back projection matrix P_i^* is equal to zero. Since also E_i^{MF} is equal to zero, the resistivities $Y_{i+1}^* = \left(P_i^{*T} P_i^*\right)^{-1} P_i^{*T} E_i^{MF}$ are undetermined, like the quotient 0/0 is undetermined. Denoting small inaccuracies in measured heads by ε, we may then find results like the quotients $0/\varepsilon$ or $\varepsilon/0$. Such instabilities can be overcome by an approach presented in Chap. 5, Eq. 5.1, as well as by zonation, as presented in Chap. 6. This type of argument against direct inversion, often presented in a more sophisticated way, shows that for small head gradients the solution becomes unstable. However, such an instability argument may not be used selectively against direct methods; it is only an argument for stabilization of the inversion method, irrespective of direct or indirect inversion. Therefore, Yeh's often-quoted demarcation between direct and indirect inversion models is not very helpful (Yeh 1986). As has been explained in Sect. 3.9.1, demarcation between calibration methods and imaging methods is more relevant.

In conclusion, constrained back projection leads to an iterative inversion method based on the flux model as forward model and a back projection model. This means that it is a calibration method, in contrast to the double constraint method, which is an imaging method because it is based on both a flux and a head model. Until now, the back projection calibration method has been applied only to a limited number of synthetic problems; the results are encouraging and, therefore, further research and development are recommended.

References

Aziz K, Settari A (1979) Petroleum reservoir simulation. Applied Science Publishers Ltd, London
Bossavit A (1998a) Computational Electromagnetism. Academic Press, Boston
Bossavit A (1998b) Computational electromagnetism and geometry. J Jpn Soc Appl Electromagn Mech 6:17–28, 114–123, 233–240, 318–326
Bossavit A (1999) Computational electromagnetism and geometry. J Jpn Soc Appl Electromagn Mech 7:150–159, 249–301, 401–408

Bossavit A (2000) Computational electromagnetism and geometry. J Jpn Soc Appl Electromagn Mech 8:102–109, 203–209, 372–377

Bossavit A (2005) Disctretization of electromagnetic problems: the "generalized finite difference approach". In: Ciarlet PG (ed) Handbook of numerical analysis, vol XIII. Elsevier, North Holland, Amsterdam

Brouwer GK, Fokker PA, Wilschut F, Zijl W (2008) A direct inverse model to determine permeability fields from pressure and flow rate measurements. Math Geosci 40(8):907–920

Butkov E (1973) Mathematical Physics. Addison Wesley Publishing Company, Reading

Chavent G (1987) On the uniqueness of local minima for general abstract non-linear least square problems. [Research report] RR-0645 <inria-00075908>. https://hal.inria.fr/inria-00075908

Chen Y, Zhang D (2006) Data assimilation for transient flow in geologic formations via ensemble Kalman Filter. Adv Water Resour 29:1107–1122

Datta-Gupta A, Yoon S, Barman I, Vasco DW (1998) Streamline-based production data integration into high resolution reservoir models. J Pet Technol 50(12):72–76

Diersch HJG (2005) FEFLOW-Finite element subsurface flow and transport simulation system. WASY GmbH, Berlin, Germany

El-Rawy M (2013) Calibration of hydraulic conductivities in groundwater flow models using the double constraint method and the Kalman filter. Ph.D. thesis, Vrije Universiteit Brussel, Brussels, Belgium

El-Rawy M, Batelaan O, Zijl W (2015) Simple hydraulic conductivity estimation by the Kalman filtered double constraint method. Ground Water 53(3):401–413. https://doi.org/10.1111/gwat.12217

Frankel T (2004) The geometry of physics, an introduction. Cambridge University Press, New York

Fry B, Wexler A (1995) US Patent 4,539,640, 3 Sept 1995

Harbaugh AW (2005) MODFLOW-2005, The US Geological Survey modular ground-water model, the ground-water flow process, techniques and methods. US Geological Survey, Reston, VA, p 6-A16

Harbaugh AW, Banta ER, Hill MC, McDonald MG (2000) MODFLOW-2000, The US Geological Survey modular ground-water model, user guide to modularization concepts and the ground-water flow process. US Geological Survey Open-File Report 00–92. Reston, Virginia: USGS, p 121

Huisman L (1950) Resistance of clay-layer Amsterdam dune water catchment area (in Dutch). Amsterdam Water Supply Report

Kaasschieter EF (1990) Preconditioned conjugate gradients and mixed-hybrid finite elements for the solution of potential flow problems. Ph.D. thesis, Delft University of Technology, Delft, Netherlands

Kaasschieter EF, Huijben SJM (1992) Mixed-hybrid finite elements and streamline computation for the potential flow problem. Numer Methods Partial Differ Equ 8:221–266

Kohn RV, McKenney A (1990) Numerical implementation of a variational method for electrical impedance tomography. Inverse Probl 6:389–414

Kohn RV, Vogelius M (1987) Relaxation of a variational method for impedance computed tomography. Comm Pure Appl Math 40:745–777

MATLAB (2017) The MathWorks, Inc., Natick, Massachusetts, USA

Mohammed GA (2009) Modeling groundwater-surface water interaction and development of an inverse groundwater modeling methodology. Ph.D. thesis, Vrije Universiteit Brussel, Brussels, Belgium. http://twws6.vub.ac.be/hydr/download/GetachewAdemMohammed.pdf

Mohammed GA, Zijl W, Batelaan O, De Smedt F (2009a) Comparison of two mathematical models for 3D groundwater flow: block-centered heads and edge-based stream functions. Trans Porous Media 79(3):469–485

Mohammed GA, Zijl W, Batelaan O, De Smedt F (2009b) 3D stream function based Hydraulic Impedance Tomography. Presented at EGU general assembly 2009, Vienna, 19–24 April 2009: 5631, session HS3.5/A108, EGU 2009, vol 11 (2009). Geophysical research abstracts, vol 11, p 5631, European Geophysical Union (2009). http://www.google.com/search?q=egu+2009 +5631+site:meetingorganizer.copernicus.org

Morse PM, Feshbach H (1953) Methods of theoretical physics. McGraw-Hill, New York

Naevdal G, Mannseth T, Vefring EH (2002) Near-well reservoir monitoring through Ensemble Kalman Filter. Paper SPE 75235 (9 pages), SPE/DOE improved oil recovery symposium, Tulsa, Oklahoma, 13–17 April 2002

Narasimhan TN (2010) The discrete and the continuous: which comes first? Curr Sci 98(8):1003–1005

Nawalany M, Zijl W (2010) The velocity oriented approach revisited. XXXVIII IAH congress, groundwater quality sustainability, extended abstract 472, topic 5: data processing in hydrogeology, 5.2: groundwater flow and solute transport modelling, Krakow, Poland, 12–17 Sept 2010. Biuletyn Państwowego Instytutu Geologicznego 441:113–122 (Geol Bull Pol Geol Inst 441:113–122)

Nelson RW (1960) In-place measurement of permeability in heterogeneous porous media 1: theory of a proposed method. J Geophys Res 65(6):1753–1758

Nelson RW (1961) In-place measurement of permeability in heterogeneous porous media 2: experimental and computational considerations. J Geophys Res 66(8):2469–2478

Nelson RW (1962) Conditions for determining areal permeability distribution by calculation. Soc Pet Eng J 2(3):223–224. https://doi.org/10.2118/371-PA

Nelson RW (1968) In-place determination of permeability distribution for heterogeneous porous media through analysis of energy dissipation. Soc Pet Eng J 8(1):33–42 http://www.onepetro.org/mslib/servlet/onepetropreview?id=00001554

Olsthoorn TN (1998) Groundwater modelling: calibration and the use of spreadsheets. Ph.D. thesis, Delft University of Technology, Delft, Netherlands

Peaceman DW (1977) Fundamentals of numerical reservoir simulation. Elsevier Science, Amsterdam

Shinbrod M (1973) Lectures on fluid mechanics. Gordon and Breach, New York

Tamburi A, Roeper U, Wexler A (1988) An application of impedance-computed tomography to subsurface imaging of pollution plumes. In: Collins AG, Johnson AI (eds) Ground-water contamination: field methods, ASTM Special Technical Publication 963, American Society for Testing and Materials, p 86–100

Trykozko A, Brouwer GK, Zijl W (2008) Downscaling: a complement to homogenization. Int J Num Anal Model 5:157–170

Trykozko A, Mohammed GA, Zijl W (2009) Downscaling: the inverse of upscaling. In: Conference on mathematical and computational issues in the geosciences, SIAM GS 2009, Leipzig, 15–18 June 2009

Trykozko A, Zijl W, Bossavit A (2001) Nodal and mixed finite elements for the numerical homogenization of 3D permeability. Comput Geosci 5:61–64

Van Leeuwen VCI (2005) A Matlab implementation of the edge-based face element method for forward and inverse modelling in the near-well region. M.Sc. thesis, Mathematical Institute, Utrecht University, Utrecht

Webster JG (1990) Electrical impedance tomography. Adam Hilger, Bristol

Wexler A (1988) Electrical impedivity imaging in two and three dimensions. Clin Phys Physiol Meas 9(Suppl A):29–33

Wexler A, Fry B, Neuman MR (1985) Impedivity computed tomography algorithm and system. Appl Opt 24(23):3985–3992

Yeh WWG (1986) Review of parameter identification procedures in groundwater hydrology: the inverse problem. Water Resour Res 22(2):95–108

Yorkey TJ, Webster JG (1987) A comparison of impedivity topographic reconstruction algorithms. Clin Phys Physiol Meas 8:55–62

Yorkey TJ, Webster JG, Tompkins WJ (1987) Comparing reconstruction algorithms for electrical impedance tomography. IEEE Trans Biomed Eng 34:843–852

Zijl W (2004) A direct method for the identification of the permeability field based on flux assimilation by a discrete analog of Darcy's law. Transp Porous Med 56:87–112

Zijl W (2005a) Face-centered and volume-centered discrete analogs of the exterior differential equations governing porous medium flow I: theory. Transp Porous Med 60:109–122

Zijl W (2005b) Face-centered and volume-centered discrete analogs of the exterior differential equations governing porous medium flow II: examples. Transp Porous Med 60:123–133

Zijl W (2007) Forward and inverse modeling of near-well flow using discrete edge-based vector potentials. Transp Porous Med 67:115–133

Zijl W, Mohammed GA, Batelaan O, De Smedt F (2010) Constraining methods for direct inverse modeling. XVIII international conference on water resources, CMWR, CIMNE, Barcelona. http://congress.cimne.com/cmwr2010/Proceedings/docs/p62.pdf

Zijl W, Nawalany M (2004) The edge-based face element method for 3D-stream function and flux calculations in porous media flow. Transp Porous Med 55:361–382

Zijl W, Trykozko A (2001) Numerical homogenization of the absolute permeability using the conformal-nodal and mixed-hybrid finite element method. Transp Porous Med 44:33–62

Chapter 4
Time Dependency

4.1 Storage Coefficients

Equations 2.1 and 2.2 can be written in the following way

$$-q_t + \partial_x q_x + \partial_y q_y + \partial_z q_z = 0. \quad \text{(water balance)} \quad (4.1)$$

$$\begin{pmatrix} q_t \\ q_x \\ q_y \\ q_z \end{pmatrix} = - \begin{pmatrix} k_t & 0 & 0 & 0 \\ 0 & k_x & k_{xy} & k_{xz} \\ 0 & k_{yx} & k_y & k_{yz} \\ 0 & k_{zx} & k_{zy} & k_z \end{pmatrix} \cdot \begin{pmatrix} \partial_t h \\ \partial_x h \\ \partial_y h \\ \partial_z h \end{pmatrix} \quad \text{(generalized Darcy)} \quad (4.2)$$

where 4-vector $(\partial h_t, \partial h_x, \partial h_y, \partial h_z)$ represents a short-hand notation for the generalized head gradient $(\partial h/\partial t, \partial h/\partial x, \partial h/\partial y, \partial h/\partial z)$. Equation 4.1 is a kinematic equation, i.e., an equation in which only flow rates occur. This equation is independent from the dynamic equations in which the driving force (head gradient, pressure gradient) occurs. In the water balance (Eq. 4.1) not only flux density $\mathbf{q} = (q_x, q_y, q_z)$ [L T^{-1}], but also flow rate q_t [T^{-1}] occurs. Considering an infinitely small volume $\Delta V \to 0$, quantity $q_t \Delta V$ [L^3 T^{-1}] is equal to the volume of water flowing out of this volume due to compression caused by an increase in pressure (head). The dynamic equations presented in Eq. 4.2 are Darcy's law $\mathbf{q} = -\mathbf{k} \cdot \nabla h$ relating the flux density to the spatial head gradient ∇h and the "compression law" $q_t = -k_t \partial h/\partial t$ relating the flow rate to the "temporal head gradient" $\partial h/\partial t$, where $k_t = s$ is the specific storage (see Eq. 2.1).

Under conditions explained in Chap. 2, Sect. 2.1, the off-diagonal components of "conductivity 4×4 matrix" are equal to zero. In that case, the flux model, with initial "conductivity 4-vector" $(k_{t;i}, k_{x;i}, k_{y;i}, k_{z;i})$, results in "flux 4-vector" $(q^F_{t;i}, q^F_{x;i}, q^F_{y;i}, q^F_{z;i})$ honoring the imposed flux conditions. These conditions are the fluxes known at the wells and through the boundaries, as well as the known initial fluxes. To be sure, at boundary locations where the fluxes are not known we impose

© The Author(s) 2018
W. Zijl et al., *The Double Constraint Inversion Methodology*,
SpringerBriefs in Applied Sciences and Technology,
https://doi.org/10.1007/978-3-319-71342-7_4

another condition (head condition or a mixed head–flux condition). In addition, if the initial flux condition is unknown, as is generally the case, we impose another initial condition.

The head model, again with "conductivity 4-vector" $(k_{t;i}, k_{x;i}, k_{y;i}, k_{z;i})$, results in head h_i^H and head gradient 4-vector $(e_t^H, e_x^H, e_y^H, e_z^H) = -(\partial_t h_i^H, \partial_x h_i^H, \partial_y h_i^H, \partial_z h_i^H)$ honoring the imposed head conditions (including the heads measured in the observation wells). Again, at boundary locations where the heads are not known we impose another condition. If the initial head conditions are unknown, as is generally the case, we impose the same condition as used for the flux model (for instance, a guess of the initial head).

In addition to Eq. 3.8, we honor Eq. 4.2 by introducing a new specific storage $k_{t+i+1} = |q_{t;i}^F / e_{t;i}^H|$. Substituting $q_{t;i}^F = k_{t+i} e_{t;i}^F$ results in

$$k_{t;i+1} = k_{t;i} \frac{|q_{t;i}^F| + q_t^\varepsilon}{|q_{t;i}^H| + q_t^\varepsilon} \tag{4.3}$$

where $q_t^\varepsilon > 0$ is a small flow rate to account for incompressible flow $(\partial_x q_x + \partial_y q_y + \partial_z q_z = 0)$. In almost all practical applications dealing with shallow groundwater, the flow may be considered as incompressible; in such cases, update rule (4.3) yields the trivial result $k_{t;i+1} = k_{t;i}$. However, update rule (4.3) plays a role in pumping well tests set up in such a way that compressibility (the term $q_t = -k_t \partial h / \partial t$ in Eq. 4.1) is not negligible (De Smedt et al. 2017).

Even if update rule Eq. 4.3 does not play a role (incompressible flow), the specific yield may play a role. Boundary condition Eq. 2.5 on the water table $z = \zeta(x, y, t)$ can be written as

$$-q_\theta + q_x \frac{\partial \zeta}{\partial x} + q_y \frac{\partial \zeta}{\partial y} - q_z = r_{eff} \tag{4.4}$$

$$q_\theta = -k_\theta \frac{\partial \zeta}{\partial t} \tag{4.5}$$

in which $k_\theta = \theta$ [–] is the specific yield. In Eq. 4.4, only flux densities occur, independent from the "Darcy-like" phenomenological equation, Eq. 4.5, which is in fact an approximation of the interaction between the saturated groundwater domain and the unsaturated zone. In Eq. 4.4 not only flux density $\mathbf{q}$ [L T^{-1}] occurs, but also flux density q_ϑ [L T^{-1}]. The recharge rate r_{eff} of water flowing from above (from the unsaturated zone) upon the water table is separated into two parts: (i) Darcy flux $q_x \partial \zeta / \partial x + q_y \partial \zeta / \partial y - q_z$ flowing through the water table into the saturated groundwater domain and (ii) accretion flux $-q_\theta$ causing the water table to rise (increasing the volume of water in the saturated domain) with rate $\partial \zeta / \partial t = -q_\theta / k_\theta$ (Eq. 4.5).

The head model with specified initial specific yield θ_i yields the water table height ζ_i^H and its rate of increase $\partial \zeta_i^H / \partial t$. The flux model, again with initial θ_i, yields accretion flux $q_{\theta;i}^F$. Requiring that the accretion flux obtained by the flux model results in the rate of water table rise $\partial \zeta_i^H / \partial t$ obtained by the head model, the initial specific yield has to be replaced with a new one that honors Eq. 4.5. This results in the same equation as Eq. 4.3; only subscript t has to be replaced with subscript θ.

In many practical applications, dealing with shallow groundwater, the flow has to be considered as time-dependent. In a number of such cases, update rule (4.3) may play a role for the specific yield k_θ. However, in quite a number of cases, Eqs. 4.4 and 4.5 are linearized and averaged over a longer time, over the interval from time t_0 to time t_1. Defining the flux averages $\bar{q}_z = (t_1 - t_0)^{-1} \int_{t_0}^{t_1} q_z dt$ and $\bar{r}_{eff} = (t_1 - t_0)^{-1} \int_{t_0}^{t_1} r_{eff} dt$, the linearized Eqs. 4.4 and 4.5 result in

$$k_\theta \frac{\zeta(t_1) - \zeta(t_2)}{t_1 - t_2} - \bar{q}_z = \bar{r}_{eff} \tag{4.6}$$

In such applications, the averaged time-dependent term $\zeta(t_1) - \zeta(t_2)/(t_1 - t_2)$ is negligibly small with respect to the average fluxes, i.e., $\bar{q}_z \approx -\bar{r}_{eff}$, which makes the flow problem quasi-steady, thus reducing Eq. 4.3 to a trivial update. Nevertheless, update rule (4.3) will always play a role in pumping well tests, because in such tests phreatic storage term $q_\theta = -k_\theta \partial \zeta / \partial t$ is never negligible.

4.2 Observation Error Versus Estimation Uncertainty: The Linear Kalman Filter

Many hydrogeological studies are devoted to the evolution in time of the groundwater heads and fluxes. Although the parameters are time-independent, estimation of these parameters by the double constraint method will generally show a time evolution of the parameters too. This parameter evolution is caused by the inaccuracies in the measured heads and fluxes that are imposed as boundary conditions. The N spatially distributed conductivities $k_{n;m}^o$ ($n = 1, 2, \ldots, N$) "observed" at time t_m by the double constraint method can be time-averaged to find the time-independent conductivities $\bar{k}_n^o$. However, to perform an analysis based on Gaussian statistics, we introduce the observed log-conductivities $z_{n;m}^o = \ln k_{n;m}^o$. Averaging the time series of conductivity values obtained for conductivity n over a time interval containing M observations $m = 1, 2, \ldots, M$ results in the time-independent observed log-conductivity n

$$\bar{z}_n^o = \frac{1}{M} \sum_{m=1}^{M} z_{n;m}^o \tag{4.7}$$

The observation error, or unconditional uncertainty, is then

$$s_n^o = \sqrt{\frac{1}{M-1}(z_{n;m}^o - \bar{z}_n^o)^2} \tag{4.8}$$

According to Bayesian probability theory, the observation uncertainty will be less than the observation error given by Eq. 4.8. In Bayesian theory, the conditional probability, denoted as $P(A|B)$, plays a central role. It is the probability of observing event A given that event B is true. Bayes's theorem states that $P(A|B) = cP(A)$ with $c = P(B|A)/P(B)$, where $P(A)$ and $P(B)$ are the probabilities of observing events A and B without regard to each other. Using the expression for c, it can be shown that for events related by an underlying process $c > 1$. In our problem dealing with parameter estimation, a probability may be considered as a measure for the degree of certainty. Let us consider $P(A|B)$ as the probability of the parameters observed at time t_{m+1} (event A) taking into account the parameters observed at an earlier time t_m (event B), while $P(A)$ is the probability of event A when event B would not have taken place. In other words, $P(A|B)$ represents the degree of certainty at observation time t_{m+1} while $P(A)$ represents the degree of certainty at the earlier time t_m. And because $c > 1$, the degree of uncertainty $1 - P(A|B)$ at time t_{m+1} is smaller than the degree of uncertainty $1 - P(A)$ at time t_m. The more observations become available, the more we may trust the time average obtained from Eq. 4.7; the results become much less uncertain than expressed by Eq. 4.8. For application of history matching by Bayesian estimation see, for instance, Gavalas et al. (1976), and for more details on probability theory see Jaynes (2003).

The above-presented ideas can be worked out using the Kalman filter, in which Bayesian probability theory is embedded. The Kalman filter is applicable under the assumption that the observations, with error determined by Eq. 4.8, are realizations of serially uncorrelated random observations with zero mean and finite variance (white noise) with Gaussian statistics. A detailed explanation of the Kalman filter, both for linear and nonlinear processes, is presented in Sect. 4.3; below we present only the results derived in Sect. 4.3.1 from the linear Kalman filter theory. As a result, we consider the following simple recurrence relations for the analysis state $x_{n;m}^a$ of log-conductivity n after observation m

$$x_{n;m}^a = g_m z_{n;m}^o + (1 - g_m)x_{n;m-1}^a \tag{4.9}$$

$$\sigma_{n;m} = \sqrt{g_m}\, s_n^o \tag{4.10}$$

$$g_m = \frac{g_{m-1} + b}{1 + g_{m-1} + b}, \quad g_1 = 1 \tag{4.11}$$

where standard deviation $\sigma_{n;m}$ represents the uncertainty in the analyzed log-conductivity $x_{n;m}^a$; g_m is the Kalman gain after observation m while dimensionless quantity b represents the relative model error.

According to Darcy's law, the model "conductivity at time m equals conductivity at time $m - 1$" is correct for the groundwater flow as it occurs in reality. However, in our approach, we do not deal with the real groundwater flow; we deal with the flow as calculated by the flux model and the head model. These models do not represent reality exactly, but only approximately. In the double constraint method, the conductivities for the grid volumes (or for larger units; see Chap. 6) are estimated by equating them to the flux calculated from the flux model divided by the head gradient calculated by the head model. Depending on the reliability of the flux and head model, this flux/head gradient ratio—i.e., the model's conductivity—may vary somewhat when the flow pattern is changing in time. As an example, we illustrate this idea by considering models in which the conductivities are assumed to be isotropic. In that case, conductivity $k_{n;i+1}$ in grid volume n is determined from the calculated flux/head gradient ratio $|q^F_{n;i}|/|\nabla h^H_{n;i}|$ (Chap. 3, Sect. 3.6). If the real conductivity is not exactly isotropic, the thus-determined conductivity will turn out to be dependent on the time-dependent flow pattern. For instance, the calculated $k_{n;i+1}$ will be smaller for flow patterns in which the flow is more vertical and will be larger for flow patterns in which the flow is more horizontal. In the Kalman filter, such deviations from the model "conductivity at time $m + 1$ equals conductivity at time m" can be accounted for by the introduction of a finite model quality $0 < 1/b < \infty$, which causes that estimation uncertainty $\sigma_{n;m}$ does not tend to zero when the time series grows longer and longer. Some uncertainty will always be present.

The Kalman filter is based on the assumption that the log-conductivities $z^o_{n;m}$ observed by the double constraint method (DCM) may be considered as a Gaussian random process. In addition, the Kalman filter requires a sequence (history) of independent observations. As a consequence, the iterative DCM update procedure $(k_{n;i+1})_m = (k_{n;i}|q^F_{n;i}/q^H_{n;i}|)_m$ (described in Chap. 3) has to be based on independently chosen initial conductivities $(k_{n;1})_m$. In other words, at observation time m conductivity $(k_{n;1})_m$ may not be influenced by knowledge obtained from the earlier DCM observations 1, 2, ..., $m - 2$, $m - 1$. The simplest way to obtain this independency is to specify the same initial conductivities $k_{n;1}$ for all observation times m. In that case, only the observation errors and uncertainties in the measurement ranges are meaningful; the zero, or almost zero, observation errors and uncertainties in the terra incognita are meaningless (see Chap. 3, Sect. 3.2). In other words, s^o_n and $\sigma_{n;m}$ determined by Eqs. 4.1 and 4.10 are meaningful only in the measurement ranges and have to be replaced with a realistic assessment of the observation errors. Such a realistic assessment can be obtained by choosing, at each observation time m, different initial conductivities $(k_{n;1})_m$ obtained from a number of realistic alternative geological models (for instance, based on geological or hydrogeological "soft data"). In the Kleine Nete case study presented in Chap. 5, these two approaches have been applied and compared.

Let us, for the moment, assume that the model is error-free; i.e., its quality factor $1/b$ is infinitely large. Then recurrence equation Eq. 4.11 yields the following result for the Kalman gain

$$g_m = \frac{1}{m} \tag{4.12}$$

Substitution into Eq. 4.9 yields

$$x_{n;M}^a = \frac{1}{M} \sum_{m=1}^{M} z_{n;m}^o = \bar{z}_n^o \tag{4.13}$$

while Eq. 4.10 results in

$$\sigma_{n;M} = \frac{1}{\sqrt{M}} s_n^o \tag{4.14}$$

We observe from Eq. 4.13 that without model error the Kalman estimate is equal to the moving average of the observations, from which it follows that in that case Eq. 4.7 is the correct estimate. More importantly, Eq. 4.14 shows that the estimation uncertainty decreases below the observation error presented by Eq. 4.8 and will eventually become zero if there were no model error. The thus-obtained simple Eq. 4.14 is well known in metrology and is presented in many texts on measurement techniques for industrial engineering to show the difference between measurement error and uncertainty; see for instance, Bell (2001).

Equation 4.13 suggests that the Kalman estimate is independent from the order in which the observations have taken place. For instance, for 100 observations ($M = 100$), the Kalman estimate $x_{n;100}^a$ for the observed time series $z_{n;1}^o, z_{n;2}^o, z_{n;3}^o, \cdots, z_{n;99}^o, z_{n;100}^o$ is the same as for the observed time series $z_{n;100}^o, z_{n;2}^o, z_{n;99}^o, \cdots, z_{n;2}^o, z_{n;1}^o$. This principle has been applied in the Kleine Nete case study presented in Chap. 5.

However, in reality, it is not that simple. The exact Kalman filter recurrence equations show that, for a nonzero model error $b > 0$, the above approximation looses its validity after a number of, say $M_\infty < \infty$, observations (see below). After these M_∞ observations, the Kalman filter becomes steady (see Sect. 4.3.3 of this chapter). Then its estimate is no longer exactly equal to the moving arithmetic average (Eqs. 4.7 and 4.13), but to a weighted average in which the last observations have greater weight than the earlier observations. However, such theoretical differences do not have practical consequences; also when a model error is accounted for, Eqs. 4.7 and 4.13 may be applied as a good approximation, as has been verified by numerical experiments based on the Kleine Nete case study (Chap. 5).

More importantly, after these M_∞ observations, the estimation uncertainty does no longer decrease; i.e., in Eq. 4.14 the number M has to be replaced with M_∞. The observation sequence (observation history) may then be terminated because more reliable calibration results cannot be obtained.

For a sufficiently high-quality factor $1/b$ (e.g., $1/b \geq 1000$) Eq. 4.11 shows that, for sufficiently small m, $g_m = 1/m$ and $\sigma_{n;m} = s_n^o/\sqrt{m}$, which shows that the observation uncertainty decreases below the observation error. However, when m becomes sufficiently large, Eq. 4.11 shows that the Kalman filter becomes steady, i.e., $g_m \to g_\infty \approx \sqrt{b}$ independent from further recursion steps (for a proof see Sect. 4.3.3 of this chapter). As a consequence, after a sufficient number of Kalman filter steps M_∞ the situation is reached where $g_\infty \approx 1/M_\infty$ and $g_\infty \approx \sqrt{b}$; that is, we find

$$M_\infty \approx \sqrt{\frac{1}{b}} \qquad (4.15)$$

For instance, assuming a quality factor $1/b \approx 1000$, we find for Eq. 4.10 after $M_\infty = 32 = \sqrt{1024}$ recurrence steps based on 32 log-conductivity observations that the estimation uncertainty is equal to $\sigma_{n;M_\infty} = s_n^o \sqrt[4]{b} \approx 0.18 \times s_n^o$. In other words, the uncertainty does no longer decrease: the Kalman filter recurrence may be terminated because more reliable results cannot be obtained.

From the above discussion, we see that an estimate of model quality $1/b$ (i.e., relative model error b) is, in fact, an estimate of the smallest possible observation uncertainty with respect to the observation error. This estimate is more or less subjective and depends on the experts' experience in confrontations between stubborn reality with results obtained from modeling and its underlying assumptions and soft data. For this estimation, we have also to take into account that the Kalman filter should be able to react to a trend reversal, i.e., to an unanticipated change in the hydrogeological situation, for instance, a relatively sudden change in conductivity caused by subsidence/compaction or an earthquake. This means that in the limit of a great many of observations the Kalman gain has to be finite ($g_\infty > 0$). Indeed, if this limit tends to $g_\infty \to 0$ Eq. 4.9 results in $x_{n;m}^a \to x_{n;m-1}^a$; a situation in which the log-conductivities are fully determined by the model and cannot, therefore, react to the observations $z_{n;m}^o$. More generally, a steady Kalman filter with nonzero Kalman gain is especially important when the Kalman filter does not only handle the parameters, but also other state variables like heads and fluxes. In such applications, the Kalman filter is not terminated, like in the above-presented example, but it is continued with a constant, observation-independent Kalman gain: the so-called steady Kalman filter (SKF); also see Sect. 4.3.3. In conclusion, we may not trust a model for 100%. There has to be room for doubts, for assimilation of additional observations. This can be accomplished by a finite steady Kalman gain (also see Sect. 4.3.3).

Quite a number of modeling studies are devoted to flow in which not only specific storage is negligible (i.e., incompressible flow) but in which also the phreatic storage term in Eq. 4.6 is negligibly small. Such studies are generally based on time-averaged models and time-averaged flux and head measurements (see Eq. 4.6). Assuming that the time-averaged model is approximately linear, we may apply Eqs. 4.13 and 4.14 to determine the log-conductivities and the observation uncertainty with respect to the observation error, $\sigma_{n;M}/s_n^o$. However, in this type of time-averaged modeling, the error cannot be determined by Eq. 4.8. This idea has been illustrated in the Schietveld case study presented in Chap. 5.

4.3   In Depth: A Closer Look at the Kalman Filter

In contrast to Sect. 4.2, this section briefly introduces the Kalman filter in its general form. A complete description of the Kalman filter can be found in Jazwinski (1970) and Maybeck (1979) while a good introduction is presented by Welch and Bishop (2006). Below we present an introduction based on the line of thought developed at Deltares (Delft, The Netherlands) in the context of oceanography; see, for instance, Zijl et al. (2015). While keeping their line of thought, we have rewritten their original text in order to address the specific problems encountered in porous media flow.

4.3.1   The Kalman Filter (KF)

This section deals with the original Kalman filter, initially introduced by Kalman (1960). Generally speaking, the Kalman filter is a recursive algorithm for estimating the state of a time-dependent system. The estimation is accomplished by a linear combination of spatially distributed observations and a forecast (a prediction by a forecast model). To be sure, the three concepts "state," "observation", and "model" introduced in the terminology of the Kalman filter theory do not necessarily correspond with the meaning of these concepts in physics, hydrogeology, or petroleum reservoir engineering.

In a groundwater flow system (an aquifer–aquitard system), the N_S components $x_{n;m}$ ($n = 1, 2, \ldots, N_S$) of the Kalman filter's state vector $\underline{x}_m$ at time t_m are the heads in the grid volumes, as well as the conductivities in the grid volumes (in some cases also the specific storage coefficients in the grid volumes and the specific yields on the water table). The Kalman filter's forecast model consists of the following two hydrogeological models:

(i)   The parameter forecast model. This model simply states that the value of a parameter (conductivity, specific storage coefficient, specific yield) at observation time t_m is equal to its value at an earlier observation time $t_{m-1} < t_m$.

(ii) The head and flux forecast model. Conventionally this model is the flux model, as defined in Chaps. 2 and 3.

Since the measured fluxes (pumping well rates, recharge rates) are already embedded in the Kalman filter's forecast model, these fluxes are not considered as observations in the Kalman filter sense. In contrast to conventional hydrogeological applications of the Kalman filter, we propose to extend the Kalman filter observations by considering some parameters as "observed" parameters in the Kalman filter sense. In this unconventional extension, these observations may come from different sources, for instance, from laboratory measurements on borehole permeabilities, from extrapolations of outcrop data, from geological rock models, in summary, from "soft data." Here, we will mainly focus on conductivities "observed" by the double constraint method. As a consequence, in this approach, the N_O components $z_{n';m}$, $n' = 1, 2, \ldots, N_O$, of observation vector $\underline{z}_m$ at time m are not only heads, but may also be conductivities.

The state estimation is performed at each time t_m where observation m comes available. The update of state vector $\underline{x}_{m-1}$ (a column vector with N_S components) at observation time $m - 1$ to state vector $\underline{x}_m$ at observation time m is accomplished by the stochastic forecast equation

$$\begin{array}{ll} \text{General KF} & \text{KF for simple model} \\ \underline{x}_m = F_m \underline{x}_{m-1} + \underline{w}_m & \underline{x}_m = \underline{x}_{m-1} + \underline{w}_m \end{array} \qquad (4.16)$$

where "General KF" shows the general Kalman filter, while "KF for simple model" shows the Kalman filter in which only conductivities, not heads, are contained in the state vector and in which all state vector conductivities are "observed" (in this case $N_O = N_S$). Vector $F_m \underline{x}_{m-1}$ (a column vector with N_S components) represents the forecast model, while vector $\underline{w}_m$ accounts for the random forecast errors (not for the systematic errors) by Gaussian white noise with zero ensemble mean. To keep all state variables and observations within the interval $-\infty < x_{n;m} < \infty$, as required for a Gaussian distribution, we do not use conductivity $k_{n;m}$ in vector $\underline{x}$, but we use instead its logarithm $x_{n;m} = \ln k_{n;m}$. For the specific storage coefficient and the specific yield, we may use $x_{n;m} = \ln s_{n;m}$ and $x_{n;m} = \ln[(\theta_{n;m}/(1 - \theta_{n;m})]$, respectively; these transformation also guarantee that the parameters remain nonnegative.

Matrix notation $F_m \underline{x}_{m-1}$ for the forecast model suggests that matrix F_m (an $N_S \times N_S$ matrix) is independent from the state vector. That is, this notation suggests that the forecast of time m is a linear combination of the state vector components at earlier observation time $m - 1$. In the original filter as proposed by Kalman (1960), this was indeed the case. However, in models for groundwater flow or petroleum reservoir simulations, matrix F_m generally depends on the state vector. Let us, for instance, consider a simple flux model for quasi-steady-state groundwater flow (Eq. 2.3) and specified conductivities (only heads in the state vector). Under these conditions, the groundwater flow model to calculate the heads $\underline{x}_m$ (Eq. 2.4) results in a forecast model $F_m \underline{x}_{m-1}$ that is independent from the state vector (i.e., matrix F_m depends on $x_{n;m-1}^{-1}$). In that case, forecast model $F_m \underline{x}_{m-1}$ depends only on time-dependent data like recharge rates at time m.

The above update (Eq. 4.16) is related to the N_O observations $\underline{z}_m$ (a column vector or $N_O \times 1$ matrix) by the stochastic observation equation

$$\text{General KF} \qquad \text{KF for simple model}$$
$$\underline{z}_m = H_m \underline{x}_m + \underline{v}_m \qquad \underline{z}_m = \underline{x}_m + \underline{v}_m \tag{4.17}$$

where $N_O \times N_S$ matrix H_m is the observation matrix. Vector $\underline{v}_m$ (an $N_O \times 1$ matrix) represents white noise with zero ensemble mean; it accounts for the random errors in the observations (again, not for the systematic errors).

In the Kalman filter, the random errors are assumed to be Gaussian and independent from observation $m - 1$ to observation m. These errors are represented by covariance matrices Q_m (an $N_S \times N_S$ matrix) for model noise $\underline{w}_m$ and R_m (an $N_O \times N_O$ matrix) for observation noise $\underline{v}_m$.

The Kalman filter results in an analysis state $\underline{x}_m^a$. This analysis state may be considered as a weighted average between a forecast state $\underline{x}_m^f$ obtained by the forecast model and the observation state $\underline{z}_m^o$ obtained from the observations. The analysis state is obtained by the forecast step followed by the analysis step.

In the forecast step, forecast mean vector $\underline{x}_m^f$ and its covariance matrix P_m^f (an $N_S \times N_S$ matrix) are updated by the forecast model (the groundwater flow model) resulting in (see Eq. 4.16)

$$\text{General KF} \qquad \text{KF for simple model}$$
$$\underline{x}_m^f = F_m \underline{x}_{m-1}^a \qquad \underline{x}_m^f = \underline{x}_{m-1}^a \tag{4.18}$$

$$P_m^f = F_m P_{m-1}^a F_m^T + Q_m \qquad P_m^f = P_{m-1}^a + Q_m \tag{4.19}$$

In which superscript T denotes the transpose of the matrix.

In the analysis step, the forecast state is combined linearly with the available observation state $\underline{z}_m^o$ to obtain the analysis state $\underline{x}_m^a$

$$\text{General KF} \qquad\qquad \text{KF for simple model}$$
$$\underline{x}_m^a = \underline{x}_m^f + G_m(\underline{z}_m^o - H_m \underline{x}_m^f) \qquad \underline{x}_m^a = \underline{x}_m^f + G_m(\underline{z}_m^o - \underline{x}_m^f) \tag{4.20}$$

$$P_m^a = (I - G_m H_m) P_m^f \qquad P_m^a = (I - G_m) P_m^f \tag{4.21}$$

where the Kalman gain matrix G_m (an $N_S \times N_O$ matrix) is given by

$$\text{General KF} \qquad\qquad \text{KF for simple model}$$
$$G_m = P_m^f H_m^T (H_m P_m^f H_m^T + R_m)^{-1} \qquad G_m = P_m^f (P_m^f + R_m)^{-1} \tag{4.22}$$

and I is the unit $N_S \times N_S$ matrix. Instead of denoting the Kalman gain by the usual symbol K, we denote it by the symbol G to avoid confusion with the hydraulic conductivity. The expressions for the optimal analysis state, the state in which we are interested, can be obtained by elimination of the forecast state vector $\underline{x}_m^f$ and the forecast covariance matrix P_m^f.

For the simple model, "parameter value at observation m is equal to parameter value at observation $m - 1$," $N_S = N_O$ and the resulting equations for the analysis state are

$$G_m = (P_{m-1}^a + Q_m)(P_{m-1}^a + Q_m + R_m)^{-1} \tag{4.23}$$

$$\underline{x}_m^a = G_m \underline{z}_m^o + (I - G_m)\underline{x}_{m-1}^a \tag{4.24}$$

$$P_m^a = (I - G_m)(P_{m-1}^a + Q_m) \tag{4.25}$$

In these equations, R_m is the covariance matrix of the observation error while Q_m is the covariance matrix of the model error (forecast error). If the flow system is stable and time-invariant, these matrices do not change much during the recursion time interval from time $m = 1$ to time $m = M$. Under the condition that they remain unchanged, we may omit index m in observation error R and in model error Q. It is then reasonable to assume that these matrices are proportional to each other, i.e., we assume $Q = bR$ in which constant b represents the model error with respect to the observation error (see Sect. 4.2 for the justification of relating model error to observation error). Equation 4.25 shows that the analysis covariance matrix of the uncertainty after analysis step, m, P_m^a, changes during the recursion. Because the Kalman filter incorporates Bayesian probability calculus, we may expect that the analysis uncertainty is initially (at time $m = 1$) equal to the observation error and decreases gradually when more and more observations come available (at times $1 < m \le M$). These assumptions result in $P_{m-1}^a = g_{m-1}R$ and substitution into Eqs. 4.23, 4.24, and 4.25 yields

$$G_m = g_m I = \frac{g_{m-1} + b}{1 + g_{m-1} + b} I \tag{4.26}$$

$$\underline{x}_m^a = g_m \underline{z}_m^o + (1 - g_m)\underline{x}_{m-1}^a \tag{4.27}$$

$$P_m^a = g_m R \tag{4.28}$$

From Eq. 4.26, we see that $g_m < g_{m-1} < 1$, which clearly illustrates the well-known engineering wisdom: uncertainty is not equal to measurement error. Equations 4.26, 4.27, and 4.28 form the basis of the uncertainty/error analysis presented in this chapter, Sect. 4.2.

The Kalman filter results in a state estimate $\underline{x}_m^a$ that is optimal in various senses: minimum variance, maximum likelihood, and mean square error senses (Maybeck 1979). However, some practical difficulties hamper its implementation for large models (e.g., hydrogeological, reservoir engineering, meteorological, oceanographic models).

The simple case "parameter value at observation m is equal to parameter value at observation $m - 1$" is computationally cheap and the assumption of linearity upon which the Kalman filter is based is satisfied (because $F_m = I$ is the unit matrix).

However, the general Kalman filter is difficult to apply because of computational cost to propagate the forecast and its error covariance. For a system with N_S state variables, covariance P^a is an $N_S \times N_S$ matrix. A typical groundwater flow model has often hundreds of thousands of variables (heads and conductivities in the grid volumes). Hence, it is computationally too costly to implement the Kalman filter in its original form. In addition, the linearity assumption of the forecast model is not honored; a groundwater flow model combined with parameter identification is nonlinear. In the next two sections, two approaches that can solve these issues are briefly summarized.

4.3.2 The Ensemble Kalman Filter (EnKF)

Although this book does not present a case study based on the ensemble Kalman filter (EnKF), we introduce this method because the EnKF is a promising method gaining more and more popularity in all branches of science and technology dealing with forecasts. Moreover, the reinvention of the DCM by Brouwer et al. (2008) was intended as a remedy for the loss of geological information caused by EnKF-smoothing (see Sect. 4.3.4). The EnKF was originally introduced by Evensen (1994) for numerical weather forecasting, and the method is also popular in oceanography (Zijl et al. 2015), hydrology (Chen and Zhang 2006), and petroleum reservoir engineering (Naevdal et al. 2002; Aanonsen et al. 2009). Below we present an introduction to the EnKF that was originally written for oceanographic applications (Zijl et al. 2015) and has substantially been modified by us to be applicable for hydrology and petroleum reservoir engineering.

The EnKF is a Monte Carlo approximation of the linear Kalman filter introduced in Sect. 4.3.1. In the original Kalman filter (KF), the probability density function (pdf) of state vector $\underline{x}_m$ is represented by an ensemble with an infinite number of ensemble members $N_R \rightarrow \infty$. However, in the EnKF, this pdf is approximated by an ensemble with a finite number of ensemble members $N_R < \infty$, for instance $N_R = 100$. As will be shown below, this discretization allows us to avoid the handling of the covariance matrix P_m (an $N_S \times N_S$ matrix), which is an extremely great advantage, because in almost all practical applications of the Kalman filter, the number of state variables N_S (in hydrogeological applications the heads and the conductivities in all grid blocks) is extremely large. Another important advantage is that EnKF can handle, in a relatively simple way, models in which matrix F_m is dependent on the state vector. This allows us to apply the EnKF to all types of models, among which quasi-steady-state models, while including data assimilation (in the sense of calibration of the flux model in the strict sense as defined in Chap. 3, Sect. 3.9.1).

To deal with the EnKF, we introduce for each observation time m an ensemble matrix $X_m = (\underline{x}_{1;m}, \underline{x}_{2;m}, \dots, \underline{x}_{r;m}, \dots, \underline{x}_{N_R;m})$, an $N_S \times N_R$ matrix whose columns $\underline{x}_{r;m}$ ($r = 1, 2, \dots, N_R$) are the ensemble members. Except for the initial prior ensemble (the ensemble at observation time $m = 1$), the ensemble members are not generally independent because subsequent observation steps $m = 2, 3, \ldots$ tie the ensemble members together. However, as an approximation, we proceed with all calculations as if they were independent.

In the forecast step of the EnKF, the mean is not calculated directly like in the original Kalman filter (Eq. 4.18). Instead, the mean is determined indirectly. For each ensemble member r, forecast model $F_m \underline{x}_{r;m-1}^a$ perturbed by a realization of model error $\underline{w}_{r;m}$ is used to calculate the forecast state $\underline{x}_{r;m}^f$ at observation time m from the analysis state $\underline{x}_{r;m-1}^a$ at earlier observation time $m - 1$

$$\underline{x}_{r;m}^f = F_m \underline{x}_{r;m-1}^a + \underline{w}_{r;m} \tag{4.29}$$

Equation 4.29 can easily be used if forecast matrix F_m is dependent on state vector $\underline{x}_{r;m-1}^a$, which allows us to consider incompressible flow models. Instead of specifying the conductivities, or estimating them by DCM, the log-conductivities are included in the state vector, which results in calibration of the flux model in the strict meaning of calibration as defined in Chap. 3, Sect. 3.9.1.

The ensemble mean $\underline{x}_m^f$ is approximated by averaging overall N_R ensemble members

$$\underline{x}_m^f = \frac{1}{N_R} \sum_{r=1}^{N_R} \underline{x}_{r;m}^f \tag{4.30}$$

while the covariance matrix (an $N_S \times N_S$ matrix) is approximated by

$$P_m^f = L_m^{f;T} L_m^f \tag{4.31}$$

where L_m^f is defined as the $N_S \times N_R$ matrix

$$L_m^f = \frac{1}{\sqrt{N_R - 1}} [(\underline{x}_{1;m}^f - \underline{x}_m^f), \ldots, (\underline{x}_{r;m}^f - \underline{x}_m^f), \ldots, (\underline{x}_{N_R;m}^f - \underline{x}_m^f)] \tag{4.32}$$

In the analysis step, each ensemble member is updated by

$$\underline{x}_{r;m}^a = \underline{x}_{r;m}^f + G_m(\underline{z}_m^o - H_m \underline{x}_{r;m}^f + \underline{v}_{r;m}) \tag{4.33}$$

in which $\underline{v}_{r;m}$ is a realization of the observation error, while the Kalman gain is calculated by

$$G_m = L_m^f L_m^{f;T} H_m^T (H_m L_m^f L_m^{f;T} H_m^T + R_m)^{-1} \tag{4.34}$$

Equation 4.34 is essentially the same equation as Eq. 4.22 for the Kalman gain of the original Kalman filter. The ensemble analysis state is approximated by the mean overall ensemble members

$$\underline{x}_m^a = \frac{1}{N_R} \sum_{r=1}^{N_R} \underline{x}_{r;m}^a \tag{4.35}$$

Because the mean of infinite ensemble $\underline{v}_m$ of the observation noise is equal to zero, the mean of the finite ensemble $\underline{v}_{r;m}$ will be approximately equal to zero (at least for a sufficient number of ensemble members). Therefore, using Eq. 4.35 to determine the mean of Eq. 4.33 leads to an approximation of equation as Eq. 4.20 for the analysis state gain of the original Kalman filter. Similarly, we can use Eq. 4.21 of the original Kalman filter to find the covariance vector in the analysis state

$$P_m^a = (I - G_m H_m) L_m^{f;T} L_m^f \tag{4.36}$$

This covariance matrix is not necessary for the update of the state vector. However, it gives additional information: the variances of the uncertainties in the diagonal, $diag(P_m^a)$, of matrix P_m^a, as well as the covariances of the correlations in the off-diagonal components $P_m^a - diag(P_m^a)I$. Especially the covariances of the correlations between the log-conductivities are interesting from a hydrogeological point of view; also see Sect. 4.3.4.

In the update step (Eq. 4.33 followed by Eq. 4.35), the Kalman gain (Eq. 4.34) plays an essential role. Equation 4.34 shows that to determine the Kalman gain it is not necessary to compute the covariance matrix P_m^f from Eq. 4.31. This is in great contrast with the original Kalman filter (Sect. 3.3.1), in which the covariance matrix has first to be calculated explicitly by Eq. 4.19 and has then to be handled in Eq. 4.22 to determine the Kalman gain, which leads to excessive computational requirements, also because the matrix $H_m P_m^f H_m^T + R_m$ has to be inverted. Consequently, for problems with large state vectors and large numbers of observed data, the EnKF is the only computationally feasible approach to data assimilation compared with the original KF.

To avoid spurious correlation leading to filter divergence, which is caused by the limited ensemble size, a distance-dependent covariance localization is generally applied (Hamill et al. 2001; Houtekamer and Mitchell 1998; Lee et al. 2011; Zhang and Oliver 2011). Covariance localization cuts off longer range correlation in the error covariance at a specified distance. The accuracy of the EnKF is highly dependent on the size of the finite ensemble. To obtain sufficiently accurate results, a relatively large number of ensemble members is required, which increases the computational cost. Fortunately, for quite a number of practical applications, the computational cost can be considerably reduced by the steady Kalman filter presented in Sect. 4.3.3.

4.3.3 The Steady Kalman Filter (SKF)

As has been exemplified in Sect. 4.2, the Kalman gain G_m (Eqs. 4.23 and 4.35) of a stable time-invariant system converges for $m \rightarrow \infty$ to a finite limiting value $G_\infty \neq 0$ (Anderson and Moore 1979). This steady Kalman filter is computationally very efficient and has been proven reliable in many applications; see, for instance, Heemink (1990), Verlaan et al. (2005), Zijl et al. (2015). Different techniques have been introduced for calculating a steady Kalman gain. One method is to compute a steady Kalman gain by averaging a time series of Kalman gains computed by the EnKF (El Serafy and Mynett 2008). The resulting steady Kalman gain is then applied to the original Kalman filter equations (Eqs. 4.20 and 4.21) to determine the update of state vector $\underline{x}_m^a$ and the components of covariance matrix P_m^a in which we are interested.

As an illustration, we consider again the simple case "parameter value at observation m is equal to parameter value at observation $m - 1$" discussed in Sect. 4.3.1 and Sect. 4.2. From Eq. 4.26, it follows that $g_m g_{m-1} + g_m b - b = g_{m-1} - g_m$. In the limit $m \rightarrow \infty$ this becomes $g_\infty^2 + g_\infty b - b = 0$ resulting in $g_\infty = [(\frac{1}{2}b)^2 + b]^{1/2} - \frac{1}{2}b$. If relative model error b is small (e.g., $0 < b < 0.001$), the approximation $g_\infty \approx b^{1/2}$ is a sufficiently accurate approximation of the steady Kalman gain. From Eq. 4.28, it follows then that the vector of variances of the steady Kalman analysis state, $(\sigma_{1;\infty}^2, \ldots, \sigma_{n;\infty}^2, \ldots, \sigma_{N;\infty}^2)^T = diag(P_\infty^a)$, is equal to $b^{1/2}((s_{1;\infty}^o)^2, \ldots, (s_{n;\infty}^o)^2, \ldots, (s_{N;\infty}^o)^2)^T$, where $((s_{1;\infty}^o)^2, \ldots, (s_{n;\infty}^o)^2, \ldots, (s_{N;\infty}^o)^2)^T = diag(R)$ is the vector of variances of the observation error. In other words, the uncertainty $\sigma_{n;\infty}$ in log-conductivity $x_{n;\infty}$ as determined by the Kalman filter is a factor $b^{1/4}$ (a factor 0.2, say) times the observation error s_n^o of log-conductivity x_n "observed" by the double constraint method. In this simple case, we may terminate the Kalman filter recursion; further recursion steps do not result in a smaller uncertainty. However, when using the steady-state Kalman filter in general applications, in which not only time-independent parameters, but also time-dependent state variables like heads, flow rates play a role, continuation of the Kalman filter recursion with the SKF makes sense and is a good alternative for the computationally demanding EnKF.

4.3.4 Balancing Between Over-Smoothing and Under-Smoothing

In hydrogeological modeling and petroleum reservoir simulation, data assimilation by the ensemble Kalman filter and steady Kalman filter (EnKF/SKF) may lead to loss of heterogeneity. In such cases, the EnKF/SKF is over-smoothing; geo(hydro) logists complain that calibration causes the loss of their geological models and insights.

The first type of over-smoothing is caused by the fact that the Kalman filter is based on two-point geostatistics in the form of multi-Gaussian random fields. However, the hydrogeological system—i.e., the spatial conductivity distribution—is often better described by a multiple-point geostatistical model, especially when channels play a role in the geological structure (Huysmans et al. 2008; Huysmans and Dassargues 2009, 2011). The analysis state obtained by a Kalman update is a linear combination of the forecasted and the observed ensemble (see Eqs. 4.20 and 4.33). In this update only the variances and covariances of the forecast and the observation play a role, thereby only preserving two-point geostatistics. As a consequence, the EnKF gradually smoothes the multiple-point geostatistical model—in which more characteristics than only variance and covariance play a role—to a two-point geostatistical model. Solutions to overcome this kind of "Gaussian smoothing" have been proposed, mainly in the field of petroleum reservoir simulation (Agbalaka and Oliver 2011; Sarma and Chen 2009).

The second type of over-smoothing is caused by the fact that there are generally much less measurements than grid block conductivities that have to be calibrated. In Chap. 2, Sect. 2.4.1, it has been shown that the number of uncorrelated conductivities that can be determined is equal to the number of head and flux measurement points (minus one). The DCM in which each grid block is considered as a zone, without correlation between the conductivities in these grid block zones, leads generally to under-smoothing. On the other hand, the EnKF smoothly interpolates the ensemble means of the log-conductivities (as given by Eq. 4.35) between the measurement points. Such a smoothed conductivity field may deteriorate the initially specified geo(hydro)logical model with different rock types having appreciably different conductivities (for instance, in an aquifer-aquitard system).

This type of over-smoothing could be remediated by considering some conductivities as Kalman filter observations. For instance, in the "terra incognita" determined by the double constraint method (Chap. 3, Sect. 3.2), we could consider the conductivities obtained from one or more hydrogeological models (the soft data) as observations. This could be done in a way comparable with the double constraint approach in the case study presented in Chap. 5, Sect. 5.2. Doing so, the EnKF honors the additional geo(hydro)logical information (soft data) without deteriorating the match with the measured heads. In fact, to satisfy the geologists' need for preservation of their geological models (obtained after hard work based on their great expertise), Brouwer et al. (2008) reinvented the double constraint method as a possible remedy for the loss of geological information caused by EnKF-smoothing. This idea could be an interesting subject for future research and development. Also, see the discussion on over-smoothing in Sect. 3.5 of Chap. 3.

References

Aanonsen SI, Naevdal G, Oliver DS, Reynolds AC, Vallès B (2009) The ensemble Kalman filter in reservoir engineering: a review. SPE Journal 14(03). doi:https://doi.org/10.2118/117274-PA

Agbalaka CC, Oliver DS (2011) Joint updating of petrophysical properties and discrete facies variables from assimilating production data using the EnKF. SPE Journal 118916:318–330

Anderson BDO, Moore JB (1979) Optimal filtering. Prentice Hall, New Jersey

Bell S (2001) A beginner's guide to uncertainty of measurement. Centre for Basic, Thermal and Length Metrology, National Physical Laboratory, ISSN 1368-6550, Teddington, Middlesex, U. K. https://www.wmo.int/pages/prog/gcos/documents/gruanmanuals/UK_NPL/mgpg11.pdf

Brouwer GK, Fokker PA, Wilschut F, Zijl W (2008) A direct inverse model to determine permeability fields from pressure and flow rate measurements. Math Geosc 40(8):907–920

Chen Y, Zhang D (2006) Data assimilation for transient flow in geologic formations via ensemble Kalman Filter. Adv Water Resour 29:1107–1122

De Smedt F, Zijl W, El-Rawy M (2017) A double constraint method for analysis of a pumping test. Submitted

El Serafy GY, Mynett AE (2008) Improving the operational forecasting system of the stratified flow in Osaka Bay using an ensemble Kalman filter–based steady state Kalman filter. Water Resour Res 44(6):W06416. https://doi.org/10.1029/2006WR005412

Evensen G (1994) Sequential data assimilation with a nonlinear quasi-geostrophic model using Monte Carlo methods to forecast error statistics. J Geophys Res 99(C5):10143–10162

Gavalas GR, Shah PC, Seinfeld JH (1976) Reservoir history matching by Bayesian estimation. Soc Pet Eng 261:337–350

Hamill TM, Withaker JS, Snyder C (2001) Distance-dependent filtering of background error covariance estimates in an ensemble Kalman filter. Mon Weather Rev 129:2776–2790

Heemink A (1990) Identification of wind stress on shallow water surfaces by optimal smoothing. Stochast Hydrol Hydraulics 4:105–119

Houtekamer PL, Mitchell HL (1998) Data assimilation using an ensemble Kalman filter technique. Mon Weather Rev 126:796–811

Huysmans M, Dassargues A (2009) Application of multiple-point geostatistics on modelling groundwater flow and transport in a cross-bedded aquifer (Belgium). Hydrogeol J 17(8):1901–1911

Huysmans M, Dassargues A (2011) Direct multiple-point geostatistical simulation of edge properties for modeling thin irregularly shaped surfaces. Math Geosci 43(5):521–536

Huysmans M, Peeters L, Moermans G, Dassargues A (2008) Relating small-scale sedimentary structures and permeability in a cross-bedded aquifer. J Hydrol 361(1–2):41–51

Jaynes ET (2003) Probability theory: the logic of science. Cambridge University Press, Cambridge, UK

Jazwinski AH (1970) Stochastic processes and filtering theory. Academic Press Inc, Cambridge, UK

Kalman RE (1960) A new approach to linear filtering and prediction problems. Trans ASME. J Basic Eng 82:35–45

Lee KB, Jo GM, Choe J (2011) Improvement of ensemble Kalman Filter for improper initial ensemble. Geosyst Eng 14(2):79–84

Maybeck PS (1979) Stochastic model estimation, and control, vol 1. Academic Press Inc, Cambridge, UK

Naevdal G, Mannseth T, Vefring EH (2002) Near-well reservoir monitoring through ensemble Kalman Filter. Paper SPE 75235 (9 pages), SPE/DOE Improved oil recovery symposium, 13–17 April 2002, Tulsa, Oklahoma, USA

Sarma P, Chen WH (2009) Generalization of the ensemble Kalman Filter using kernels for non-Gaussian random fields. SPE 119177

Verlaan M, Zijderveld A, de Vries H, Kroos J (2005) Operational storm surge forecasting in the Netherlands: developments in the last decade. Phil Trans R Soc A 363:1441–1453

Welch G, Bishop G (2006) An introduction to the Kalman Filter. Technical Report 95-041, Department of Computer Science, University of North Carolina at Chapel Hill, Chapel Hill, NC 27599-3175. http://www.cs.unc.edu/~welch/media/pdf/kalman_intro.pdf

Zhang Y, Oliver D (2011) Evaluation and error analysis: Kalman gain regularization versus covariance regularization. Comput Geosci 15(3):1–20

Zijl F, Samihar J, Verlaan M (2015) Application of data assimilation for improved operational water level forecasting on the northwest European shelf and North Sea. Ocean Dyn 65 (12):1699–1716. https://doi.org/10.1007/s10236-015-0898-7

Chapter 5
Case Study Kleine Nete: Observation Error and Uncertainty

5.1 Introduction

To illustrate the difference between observation error and uncertainty, we consider a case study related to the Kleine Nete basin (El-Rawy 2013; El-Rawy et al. 2015). The basin is located in the northeast of Belgium, 65 km from Brussels, and comprises 581 km^2. The elevation of the basin is between 3 and 48 m with an average slope of 0.36%. The soil map indicates that sand is the most extensive occurring soil texture, while some loamy sand, sandy loam, and sandy clay are present in the valleys. The average precipitation is about 840 mm/y. The distributed recharge, $r_{eff}(x, y, t)$ (see top boundary conditions Eqs. 2.5–2.7), was estimated half-monthly for the year 1992 using the WetSpa model (Dams et al. 2008, 2012). The averaged recharge, about 500,000 m^3/d, is an order of magnitude larger than the 50,000 m^3/d abstracted by the 565 pumping wells. Heads were measured in observation wells, also half-monthly for the year 1992; see Fig. 5.1.

The groundwater flow model has been presented by Dams et al. (2008, 2012) and is based on MODFLOW software (Harbaugh et al. 2000; Harbaugh 2005), a flux-continuous block-centered finite difference approximation of Eqs. 2.1–2.7. The conductivities are determined by the double constraint method (DCM), which is based on two models: a flux model and a head model (Sect. 3.2). The head model matches exactly with the heads measured in the observation wells, while the, flux model matches exactly with the imposed flux data (recharge rates and pumping well rates). As a result of MODFLOW's flux-continuous approximation, the flux model over-estimates the head gradients. As a consequence, the flux model does not exactly reproduce the measured heads and cannot, therefore, be considered as a calibration method (calibration in the strict sense). In other words, the DCM is an imaging technique, as explained in Sect. 3.9 of Chap. 3). However, in the Kleine Nete model (more precisely, the Kleine Nete flux model), this head mismatch turned out to be small with respect to the measurement errors.

© The Author(s) 2018
W. Zijl et al., *The Double Constraint Inversion Methodology*,
SpringerBriefs in Applied Sciences and Technology,
https://doi.org/10.1007/978-3-319-71342-7_5

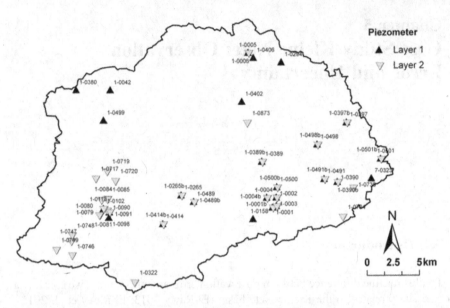

Fig. 5.1 Observation wells in Kleine Nete catchment, Belgium

The model domain is discretized horizontally in 722 columns and in 511 rows with horizontal grid block dimensions $\Delta x = \Delta y = 50$ m. In the vertical direction, the gridding consists of two layers with varying thickness $\Delta z(x, y)$. Hence, the model is discretized with 737,884 grid blocks. Provided that the vertical head gradient $\partial h / \partial z$ is small with respect to the horizontal head gradient, MODFLOW's finite difference formulation of Darcy's law yields a sufficiently accurate approximation for varying layer thickness.

The WetSpa-estimated distributed recharge $r_{eff}(x, y, t)$ gives input to MODFLOW's RCH package, while the WELL package manages the pumping wells. A no-flow condition is assumed for the catchment boundary. MODFLOW's RIVER package is used to simulate rivers, canals, and lakes. However, the DRAIN package is preferred for simulating groundwater drainage to small streams and wetlands. The drain-level parameter is estimated as the highest elevation in the soil profile where oxidation is observed. Above this level, groundwater will discharge (Batelaan and De Smedt 2004). The horizontal initial grid block conductivities (the conductivities before applying the DCM) $k_H = k_x = k_y$ are parameterized on basis of information from the hydrogeological classification system for Flanders (HCOV) (Cools et al. 2006). Model layer 1 corresponds to Quaternary, Pleistocene, and Pliocene formations, while model layer 2 represents the productive Miocene aquifer (HCOV 025).

For the conductivity estimation, update rule Eq. 3.8 is applied in the modified form.

$$k_{x;i+1} = k_{x;i} \frac{|Q^F_{x;i}| + Q^\varepsilon}{|Q^H_{x;i}| + Q^\varepsilon}$$

$$k_{y;i+1} = k_{y;i} \frac{|Q^F_{y;i}| + Q^\varepsilon}{|Q^H_{y;i}| + Q^\varepsilon} \qquad (5.1)$$

$$k_{z;i+1} = k_{z;i} \frac{|Q^F_{z;i}| + Q^\varepsilon}{|Q^H_{z;i}| + Q^\varepsilon}$$

where $Q_x = q_x \Delta y \Delta z$, $Q_y = q_y \Delta z \Delta x$, $Q_z = q_z \Delta x \Delta y$ denote the grid block averaged fluxes [$L^3 T^{-1}$] in the x, y, and z direction through the corresponding surface areas. These fluxes are obtained by linear interpolation of the face-based fluxes calculated by MODFLOW. The relatively small flux $Q^\varepsilon = 0.01$ m^3/d is introduced to avoid division by zero. Generally, division by zero may occur in grid blocks with a production/injection well (not applicable in this particular case study), at water divides, and in stagnation zones. For grid blocks in which the DCM resulted in too large anisotropy ratios, mixing rule Eqs. 3.9 and 3.10 were applied to keep anisotropy ratios within the interval $1.1 \le k_x/k_z \le 10$ while the intervals for k_y/k_z and k_x/k_y were left unspecified which leads to reasonable ratios for the Kleine Nete basin (El-Rawy et al. 2010; El-Rawy 2013).

The model has been run for incompressible flow (flow in which the specific storage term $s\,\partial h/\partial t$ in Eq. 2.4 is negligible), and also under quasi-steady conditions (phreatic storage terms $\theta\,\partial\zeta/\partial t$ and $\theta\,\partial h/\partial t$ in Eqs. 2.5 and 2.7 are negligible). The validity of these assumptions has been tested using a time-dependent model with specific storage coefficient $s = 0.5 \times 10^{-5}$ m^{-1} and specific yield $\theta = 0.25$. The differences with the incompressible and quasi-steady state model turned out to be negligible; therefore, we have chosen for the simpler and computationally much cheaper quasi-steady-state model. For more details about numerical experiments regarding the phreatic top boundary conditions, see El-Rawy (2013: 175–177).

The DCM simulations are continued till a (local) minimum difference between measured and simulated heads are obtained (also see Chap. 3, Sect. 3.7). The first local minimum is reached after a relatively small number of iterations. Considering the observation inaccuracy obtained from the time series of estimated conductivities, it is not necessary to look for a smaller local minimum, or for the global minimum. Dealing with realistic hydrogeological problems means that it makes no sense to simulate till an accuracy is obtained that is much higher than the accuracy of the uncertain head and even more uncertain flux estimates. Hence, we applied a first DCM stroke followed by two iterations.

5.2   Case 1: Observation-Independent Initial Log-Conductivities

As has been explained in Chap. 3, Sect. 3.2, and Chap. 4, Sect. 4.2, the uncertainty in the conductivities calculated by the Kalman filter equations (Eqs. 4.1 and 4.10) is meaningful only in the measurement ranges. In the terra incognita, the application of these equations is meaningless and may, therefore, be omitted from the computational scheme. However, both in case 1 presented in this section, and in case 2 presented in Sect. 5.3, we have applied these equations to all grid blocks for the purpose of demonstration.

In case 1 is for each observation time the same initial conductivities used in DCM; only the measured heads and fluxes differ at each observation time. DCM "observes" a time series of 24 different log-conductivities for each of the 737,884 grid blocks. The log-conductivity has been normalized as $x = \ln(k/k_{norm})$ with $k_{norm} = 1$ m/d. The differences between the initial conductivities (K0), the averaged DCM-observed conductivities (K_after DCM, Eq. 4.7) and the KF-estimated conductivities (K_after KF, Eq. 4.9) are shown in Figs. 5.2 and 5.3, while their standard deviations before KF (Eq. 4.8) and after KF (Eq. 4.10) are presented in Figs. 5.4 and 5.5.

It is observed that the DCM-estimated conductivities in the measurement ranges turn out to differ appreciably from the initial conductivities while the averaged and KF-estimated conductivities differ relatively slightly, which shows that Eq. 4.13 is a reasonable approximation, at least for this case with model error $b = 0.001$, or model quality factor $1/b = 1000$), see Chap. 4, Sect. 4.2.

The observation errors and uncertainties (standard deviations s_n^o and $\sigma_{n;M}$, respectively) have been calculated from Eqs. 4.8 and 4.10 and are presented in

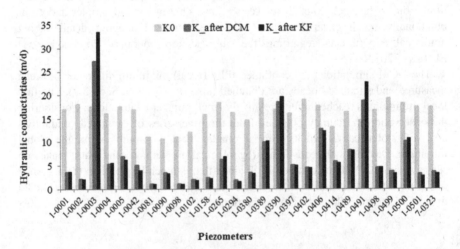

Fig. 5.2 Horizontal conductivities at piezometer locations in the top layer: K0 (light grey), after averaged DCM (dark grey) and after KF-estimation (black)

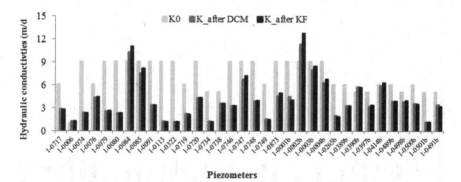

Fig. 5.3 Horizontal conductivities at piezometer locations in the bottom layer: K0 (light grey), after averaged DCM (dark grey) and after KF-estimation (black)

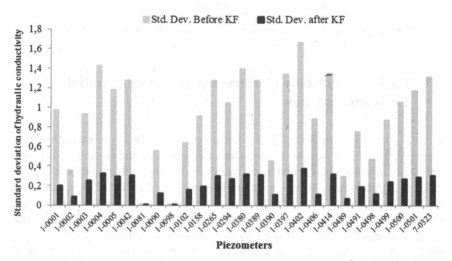

Fig. 5.4 Standard deviations of the observed conductivities at piezometer locations in the top layer before KF (grey) and after KF (black)

Figs. 5.4 and 5.5. As a result of the Kalman filter, the uncertainties are about 77% smaller than the observation errors; see Eq. 4.14 and the subsequent discussion in Chap. 4, Sect. 4.2.

Conductivities in grid blocks further away from the piezometers have (almost) zero variance and are therefore not presented; they indicate the "terra incognita," these locations are so far away from the measurements that DCM is hardly able to update these conductivities. In reality, the uncertainty of these initial conductivities will generally be much larger than the uncertainty in the measurement ranges.

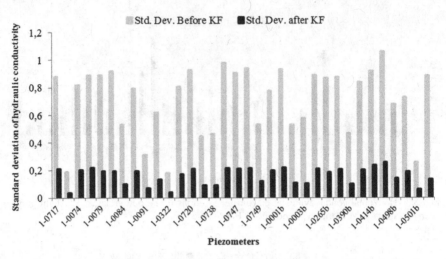

Fig. 5.5 Standard deviations of the conductivities at piezometer locations in the bottom layer before KF (grey) and after KF (black)

5.3   Case 2: Artificially Constructed Sequence of Initial Log-Conductivities

In case 1 discussed in Sect. 5.2, we keep, every time DCM estimates the conductivity, the initial conductivities the same. Consequently, the greatest observation error and uncertainty was found in the measurement ranges of the piezometers; the "terra incognita" conductivities have not been improved by DCM. Application of the Kalman filter (Eqs. 4.1 and 4.10) is meaningful only in the measurement ranges, except if the observation error in the terra incognita is specified explicitly. To show the consequences of observation errors in the initial conductivities (the "soft input data"), we introduce case 2, in which for every observation time another initial conductivity is specified. Hence, in case 2 input to DCM and the Kalman filter (KF) is, besides the measurements of the 24 different head-flux couples, also a different initial conductivity pattern for each of the 24 time sequences.

We could, for instance, consider a team (hydro) of geological experts in which each team member develops the best possible initial conductivity pattern, independent from the other team members. This ensemble of realistic initial conductivity patterns can then be used as input for the DCM and subsequent application of the KF. As has been anticipated in Chap. 4, Sect. 4.2 and shown experimentally for case 1, the order in time in which the different ensemble members are applied does not matter very much, because Eq. 4.13 is a reasonable approximation for not too large model errors. To assign realistic initial conductivities, the hydrogeologists have to make use of (hydro)geological information, as e.g., facies distributions, boreholes observations, geophysical imaging. Generally, this will be in combination with measurements of (fine-scale) conductivity from pump tests or grain size data

(Rogiers et al. 2012), which can be upscaled or homogenized (Trykozko et al. 2001; Zijl and Trykozko 2001). Information or knowledge of (conceptual) (hydro)geology is already frequently used in geostatistical methods applied in inverse modeling of groundwater flow (Doherty 2003; Doherty and Hunt 2009, 2010; Doherty et al. 2010). In this respect, one of the more advanced methods is multiple-point geostatistics, which relies on information from geological analogs (Huysmans et al. 2008; Huysmans and Dassargues 2009, 2011).

However, for the purpose of demonstration, we have used a much simpler approach than assigning a team of hydrogeologist to construct an ensemble of reasonable initial conductivity patterns. We have chosen $3N$ different initial log-conductivities x_n $(n = 1, 2, \ldots, 3N)$ of the N grid blocks (in this example $3N = 3 \times 737,884 = 2,213,652$) in such a way that the ensemble mean $\bar{x}_n$ of each log-conductivity equals the initial log-conductivity of case 1 (index n denotes the voxel number; Chap. 3, Sect. 3.3). A simple way to obtain an ensemble with members $x_{n;m}$ is to assume $x_{n;m} = [1 + c(m - (M+1)/2)]\,\bar{x}_n$, where index $m = 1, 2, \ldots, M$ is the member number and c is a constant. In our application, M is the number of observations and $x_{n;m}$ is considered as observation m of the initial log-conductivity of voxel n at time t_m (initial means before applying DCM). For these realizations, the mean is equal to $\bar{x}_{n;m}$ and is chosen equal to the initial log-conductivity applied in case 1. The difference between the two extreme values $x_{n;1}$ and $x_{n;M}$ is $x_{n;M} - x_{n;1} = c(M - 1)\,\bar{x}_n$. For sufficiently large M (theoretically for $M \rightarrow \infty$), the standard deviation of the observation error (the spread) is equal to $s_n = |x_{n;M} - x_{n;1}|/2\sqrt{3}$, which is a sufficiently accurate approximation for a finite value of M. In our example, the number of observations is equal to $M = 24$. We have chosen $c = 0.045$ resulting in $x_{n;m} = [1 + 0.045(m - 12.5)]\,\bar{x}_n$, $x_{n;1} = 0.4825\bar{x}_n$, $x_{n;24} = 1.5175\bar{x}_n$ and $\bar{x}_{n;24} - \bar{x}_{n;1} = 1.035\,\bar{x}_n$, while the spread is given as $s_n = 0.2988\,|\bar{x}_n|$. Admittedly, these realizations do not represent a Gaussian ensemble; however, as a simple method, it is sufficiently illustrative to exemplify the effect of soft data observation errors and initial conductivity uncertainties (the soft date). In Sect. 5.4, the results of a more realistic distribution are shown.

Comparison of the thus-obtained log-conductivities with the ones of case 1, with zero soft data observation error, indicates for the measurement ranges that case 1 results in slightly higher conductivities than case 2; there are small systematic differences. See Figs. 5.6 and 5.7.

The numerical experiments show that in the measurement ranges the observation errors and resulting uncertainties, in this case with appreciable soft data observation error $(c = 0.045)$, are essentially the same as in the case without zero soft data observation $(c = 0)$; the random errors (noise) have the same magnitude. See Figs. 5.8 and 5.9.

In the areas away from the measurement ranges, the observation uncertainties turn out to differ considerably. As expected, the difference is greatest in the terra incognita; i.e., the areas where the head observations cannot improve the initially selected conductivities of the double constraint method. It is also noted that both in

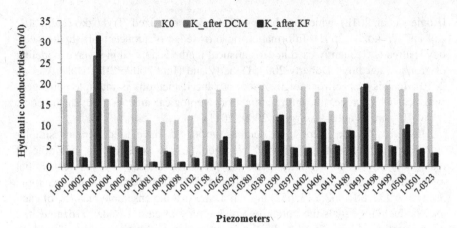

Fig. 5.6 Horizontal conductivities for case 2 at piezometer locations in the top layer: K0 (light grey), after averaged DCM (dark grey) and after KF-estimation (black)

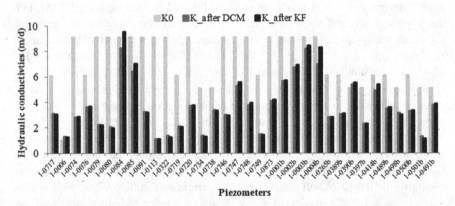

Fig. 5.7 Horizontal conductivities for case 2 at piezometer locations in the bottom layer: K0 (light grey), after averaged DCM (dark grey) and after KF-estimation (black)

the measurement ranges and in the terra incognita the uncertainty is approximately 77% smaller than the observation error.

5.4   Case 3: Realistically Constructed Sequence of Initial Log-Conductivities

The above-presented artificially constructed sequence of initial log-conductivities is useful because of its simplicity and insightful expressions for the observation error and observation uncertainty. To investigate whether the thus-obtained results make

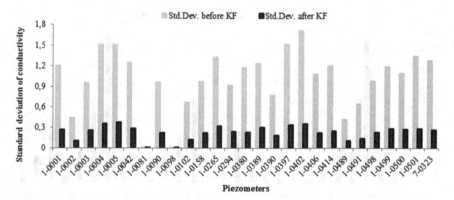

Fig. 5.8 Standard deviations for case 2 of the observed conductivities at piezometer locations in the top layer before KF (grey) and after KF (black)

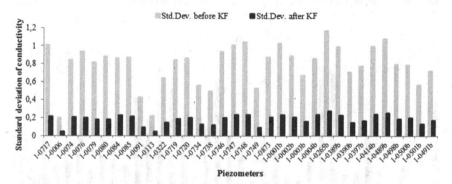

Fig. 5.9 Standard deviations for case 2 of the observed conductivities at piezometer locations in the bottom layer before (grey) and after KF (black)

sense from a more realistic point of view we have also applied 24 initial conductivities constructed by sequential Gaussian simulation with ordinary spatial kriging using the Stanford Geostatistical Modeling Software (Remy et al. 2009). The results are presented in Figs. 5.10, 5.11, 5.12 and 5.13. The figures show that, in the measurement ranges, the differences with the above-presented cases 1 and 2 are small.

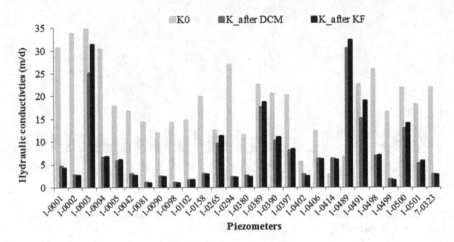

Fig. 5.10 Horizontal conductivities for case 3 at piezometer locations in the top layer: K0 (light grey), after averaged DCM (dark grey) and after KF-estimation (black)

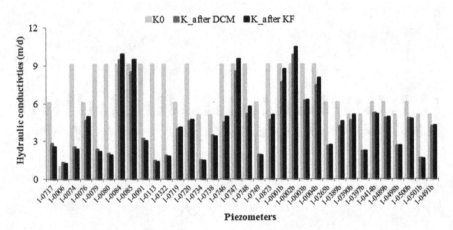

Fig. 5.11 Horizontal conductivities for case 3 at piezometer locations in the bottom layer: K0 (light grey), after averaged DCM (dark grey) and after KF-estimation (black)

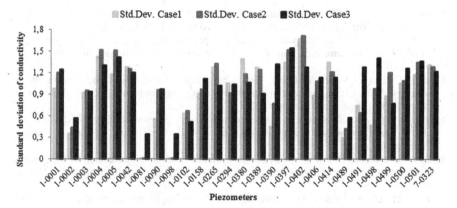

Fig. 5.12 Standard deviations for case 3 of the observed conductivities at piezometer locations in the top layer for case 1 (light-grey), case 2 (grey) and case 3 (black)

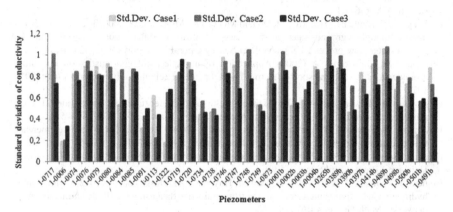

Fig. 5.13 Standard deviations for case 3 of the observed conductivities at piezometer locations in the bottom layer for case 1 (light-grey), case 2 and case 3 (black)

References

Batelaan O, De Smedt F (2004) Seepage a new MODFLOW drain package. Ground Water 42 (4):576–588

Cools J, Meyus Y, Woldeamlak ST, Batelaan O, De Smedt F (2006) Large-scale GIS-based hydrogeological modelling of Flanders: a tool for groundwater management. Environ Geol 50:1201–1209

Dams J, Woldeamlak ST, Batelaan O (2008) Predicting land-use change and its impact on the groundwater system of the Kleine Nete catchment, Belgium. Hydrol Earth Syst Sci 12:1369–1385. https://doi.org/10.5194/hess-12-1369-2008

Dams J, Salvadore E, Van Daele T, Ntegeka V, Willems P, Batelaan O (2012) Spatio-temporal impact of climate change on the groundwater system. Hydrol Earth Syst Sci 16:1517–1531

Doherty J (2003) Ground water model calibration using pilot points and regularization. Ground Water 41(2):170–177

Doherty J, Hunt RJ (2009) Two statistics for evaluating parameter identifiability and error reduction. J Hydrol 366:119–127. https://doi.org/10.1016/j.jhydrol.2008.12.018

Doherty J, Hunt RJ (2010) Response to comment on: two statistics for evaluating parameter identifiability and error reduction. J Hydrol 380:489–496. https://doi.org/10.1016/j.jhydrol.2009.10.012

Doherty JE, Hunt RJ, Tonkin MJ (2010) Approaches to highly parameterized inversion: a guide to using PEST for model-parameter and predictive-uncertainty analysis. US Geological Survey Scientific Investigations Report 2010–5211

El-Rawy M (2013) Calibration of hydraulic conductivities in groundwater flow models using the double constraint method and the Kalman filter. Ph.D. thesis, Vrije Universiteit Brussel, Brussels, Belgium

El-Rawy M, Batelaan O, Zijl W (2015) Simple hydraulic conductivity estimation by the Kalman filtered double constraint method. Ground Water 53(3):401–413. https://doi.org/10.1111/gwat.12217

El-Rawy M, Zijl W, Batelaan O, Mohammed GA (2010) Application of the double constraint method combined with MODFLOW. In: Proceedings of the Valencia IAHR International Groundwater Symposium 2010, 22–24 September

Harbaugh AW (2005) MODFLOW-2005, The U.S. Geological Survey modular ground-water model, the ground-water flow process, techniques and methods. US Geological Survey, Reston, VA, p 6–A16

Harbaugh AW, Banta ER, Hill MC, McDonald MG (2000) MODFLOW-2000, The US Geological Survey modular ground-water model, user guide to modularization concepts and the ground-water flow process. US Geological Survey Open-File Report 00-92, 121, Reston, VA

Huysmans M, Dassargues A (2009) Application of multiple-point geostatistics on modelling groundwater flow and transport in a cross-bedded aquifer (Belgium). Hydrogeol J 17(8):1901–1911

Huysmans M, Dassargues A (2011) Direct multiple-point geostatistical simulation of edge properties for modeling thin irregularly shaped surfaces. Math Geosci 43(5):521–536

Huysmans M, Peeters L, Moermans G, Dassargues A (2008) Relating small-scale sedimentary structures and permeability in a cross-bedded aquifer. J Hydrol 361(1–2):41–51

Remy N, Boucher A, Wu J (2009) Applied geostatistics with SGeMS: a user's guide. Cambridge University Press, Cambridge

Rogiers B, Mallants D, Batelaan O, Gedeon M, Huysmans M, Dassargues A (2012) Estimation of hydraulic conductivity and its uncertainty from grain-size data using GLUE and artificial neural networks. Math Geosci 44(6):739–763

Trykozko A, Zijl W, Bossavit A (2001) Nodal and mixed finite elements for the numerical homogenization of 3D permeability. Comput Geosci 5:61–64

Zijl W, Trykozko A (2001) Numerical homogenization of the absolute permeability using the conformal-nodal and mixed-hybrid finite element method. Transp Porous Med 44:33–62

Chapter 6
The Zone-Integrated Double Constraint Method

6.1 Introduction

In Chap. 2, Sect. 2.2, we have introduced smooth porous media, i.e., media for which it is required that the Laplacian of the sqrt-conductivity, $\nabla^2 \alpha$, exists in the model domain, or in a number of zones in which the model domain is partitioned. This smoothness requirement means that the conductivity value in one point cannot "jump" in an arbitrary way to a completely different value. In other words, the requirement that $\nabla^2 \alpha$ exists introduces a correlation between conductivities in neighboring points. In Sect. 2.2, it has been exemplified that, under this smoothness condition, the conductivities in the model domain can be determined uniquely by imposing heads and normal fluxes on the boundaries. This exemplification illustrates Calderón's conjecture (1980) stating that the conductivity field in the model domain is uniquely determined by head and normal flux conditions on the closed boundary, provided that the conductivity field has some smoothness or, in other words, that conductivities in neighboring points are correlated in a "smooth" way (e.g., by differentiability). The problem to determine what smoothness conditions are necessary and sufficient is generally known as the Calderón problem; also see Chap. 2, Sect. 2.4. Of course, in hydrogeological problems existence of $\nabla^2 \alpha$ is not a realistic assumption for the whole model domain but may be considered as a realistic approximation for a relatively limited number of zones within the model domain.

Application of the double constraint method in the way presented in Chap. 3 may lead to loss of smoothness at locations where smoothness is expected from a hydrogeological point of view. To avoid such a loss of smoothness, this chapter presents the simplest possible smoothness condition: constant hydraulic conductivity in a limited number of zones. Accounting for anisotropy this means that we introduce N_L conductivity zones in which the conductivities $k_{x;L}, k_{y;L}, k_{z;L}$ $(L = 1, 2, \ldots, N_L)$ are homogeneous (independent from the spatial coordinates x, y, z).

© The Author(s) 2018
W. Zijl et al., *The Double Constraint Inversion Methodology*,
SpringerBriefs in Applied Sciences and Technology,
https://doi.org/10.1007/978-3-319-71342-7_6

6.2 Minimization of Darcy Residual

In the framework of the double constraint methodology (Chap. 3), we consider the sequence of conductivities $k_{x;L;i} = \alpha_{x;L;i}^2, k_{y;L;i} = \alpha_{y;L;i}^2, k_{z;L;i} = \alpha_{z;L;i}^2$ obtained from the first DCM stroke $(i = 0)$ followed by iterations $(i = 1, 2, 3, \ldots)$ until convergence. Based on these conductivities, we calculate in the whole flow domain (including all zones) the fluxes $(q_{x;n;i}, q_{y;n;i}, q_{z;n;i})$ that honor the flux conditions (i.e., we run the flux model) and we calculate the head gradients $(e_{x;n;i}, e_{y;n;i}, e_{z;n;i})$ that honor the head conditions (i.e., we run the head model). In this notation index, $n = 1, \ldots, N$ denote the points in the model domain, as has been explained in Chap. 3, Sect. 3.3; these points represent a discrete approximation of the model domain. The limit $N \to \infty$ (N is extremely large in an unspecified way) denotes the continuum, for which powerful analytical tools based on differential and integral calculus make sense as a good approximation of reality.

To simplify the derivation, we omit the indices x, y, z and use instead the voxel notation introduced in Chap. 3, Sect. 3.3. The Darcy residual that has to be minimized with respect to $\alpha_{n;i+1}$ is (see Eq. 3.3)

$$R_{i+1} = \sum_{n=1}^{N} \left(\alpha_{n;i+1}^{-1} q_{n;i} - \alpha_{n;i+1} e_{n;i} \right)^2 \tag{6.1}$$

For the N_L, conductivities that are homogeneous in the N_L zones Eq. 3.1 can be written as

$$R_{i+1} = \sum_{L=1}^{N_L} \left(\alpha_{L;i+1}^{-2} \sum_{D_L} q_{n;i}^2 - 2N_L \sum_{D_L} q_{n;i} e_{n;i} + \alpha_{L;i+1}^2 \sum_{D_L} e_{n;i}^2 \right) \tag{6.2}$$

where D_L represents the subdomain of zone L; summation $\sum_{D_L} f_{n;i}$ is a discrete approximation of the integral $\iiint_{D_L} f_i(x, y, z) w(x, y, z) dV$ in which $dV = dxdydz$ and $w(x, y, z)$ are a suitably chosen weighting function (see Sect. 3.4). To find a minimum, we determine $\partial R_{i+1}/\partial \alpha_{L;i+1}$

$$\partial R_{i+1}/\partial \alpha_{L;i+1} = -2 \left(\alpha_{L;i+1}^{-3} \sum_{D_L} q_{n;i}^2 - \alpha_{L;i+1} \sum_{D_L} e_{n;i}^2 \right) \tag{6.3}$$

The requirement for a stationary point, $\partial R_{i+1}/\partial \alpha_{L;i+1} = 0$, results in an update rule. Going back to the notation with indices x, y, z and using volume integrals instead of summations over discrete approximations this update rule yields

$$k_{x;L;i+1} = k_{x;L;i}\sqrt{\frac{\iiint_{D_L}(q_{x;i}^F)^2 w dV}{\iiint_{D_L}(q_{x;i}^H)^2 w dV}}$$

$$k_{y;L;i+1} = k_{y;L;i}\sqrt{\frac{\iiint_{D_L}(q_{y;i}^F)^2 w dV}{\iiint_{D_L}(q_{y;i}^H)^2 w dV}} \qquad (6.4)$$

$$k_{z;L;i+1} = k_{z;L;i}\sqrt{\frac{\iiint_{D_L}(q_{z;i}^F)^2 w dV}{\iiint_{D_L}(q_{z;i}^H)^2 w dV}}$$

for each zone $L = 1,\ldots,N_L$. The second derivative of the Darcy residual is

$$\partial^2 R_{i+1}/\partial\alpha_{L;i+1}^2 = 6\alpha_{L;i+1}^{-4}\sum_{D_L}q_{n;i}^2 + 2\sum_{D_L}e_{n;i}^2 \qquad (6.5)$$

and because this second derivative (Eq. 6.5) is positive, the solution (Eq. 6.4) represents a local minimum of the residual (Eq. 6.2).

Too large anisotropy ratios may be removed by mixing rule Eq. 3.10 in which $q_{x;i+1}^F, q_{y;i+1}^F, q_{z;i+1}^F$ are replaced with $(\iiint_{D_L}(q_{x;i}^F)^2 w dV)^{1/2}, (\iiint_{D_L}(q_{y;i}^F)^2 w dV)^{1/2}, (\iiint_{D_L}(q_{z;i}^F)^2 w dV)^{1/2}$, respectively. If we want to obtain conductivities with specified anisotropy ratios, we can use Eq. 3.11 in the modified form

$$\begin{pmatrix} k_{x;L;i+1} \\ k_{y;L;i+1} \\ k_{z;L;i+1} \end{pmatrix} = \begin{pmatrix} k_{x;L;i} \\ k_{y;L;i} \\ k_{z;L;i} \end{pmatrix} \qquad (6.6)$$
$$\times \sqrt{\frac{\iiint_{D_L}[(q_{x;i}^F)^2/\eta_{xz} + (q_{y;i}^F)^2/\eta_{yz} + (q_{z;i}^F)^2]w dV}{\iiint_{D_L}[(q_{x;i}^H)^2/\eta_{xz} + (q_{y;i}^H)^2/\eta_{yz} + (q_{z;i}^H)^2]w dV}}$$

where $\eta_{xz} = k_{x;L}/k_{z;L}$ and $\eta_{yz} = k_{y;L}/k_{z;L}$ are the specified anisotropy ratios; see Chap. 3, Sect. 3.6. Also, mixed forms of dealing with anisotropy are possible. For instance, in the Schietveld case study presented in Sect. 6.4, an update rule like Eq. 6.6 has been used for the horizontal component $\iiint_{D_L}[(q_{x;i}^F)^2/\eta_{xz} + (q_{y;i}^F)^2/\eta_{yz}]w dV$ while an independent update rule has been used for the vertical component $\iiint_{D_L}(q_{z;i}^F)^2 w dV$.

In view of the approach to time-dependent flow presented in Chap. 4, Sect. 4.1, in which the generalized Darcy law $q_t = -k_t\partial h/\partial t$ is proposed, it is also possible to apply update rules like Eq. 6.6 for zones in which the zone does not consist of only a piece of space, but in which the zone contains also a piece of time. This idea has been worked out by De Smedt et al. (2017) for the interpretation of pumping well tests.

6.3   Grid Block Averaging

To show the integration for the block centered finite difference method (e.g., MODFLOW), we consider the integrals of the form $\iiint_{D_L} q_x^2 w dV$ as they occur in update rules Eqs. 6.4 and 6.6. In this method, the grid volumes are "blocks," i.e., parallelepipeds with mutually orthogonal edges in the x, y, and z directions. We denote the three coordinates of a grid block center by x_I, y_J, z_K or, in short-hand notation, by its three discretized coordinates I, J, K. The volume of each grid block is then equal to $\Delta V_{I,J,K} = \Delta x_I \Delta y_J \Delta z_K$, which makes that the integrals in Eqs. 6.4 and 6.6 can be written as

$$\iiint_{D_L} q_x^2 w dV = \sum_{(I,J,K) \in L} \iiint_{\Delta V_{I,J,K}} q_x^2 w dV \tag{6.7}$$

where $\iiint_{\Delta V_{I,J,K}} q_x^2 w dV$ is the integral over grid block (I, J, K), while $\sum_{(I,J,K) \in L}$ indicates that we have to take the sum over all grid blocks (I, J, K) within zone L.

In the mathematical assumptions underlying the mixed hybrid finite element method, or face centered finite element method, the fluxes vary linearly within a grid volume (Chavent and Jaffré 1986; Kaasschieter 1990; Kaasschieter and Huijben 1992; Zijl 2005a, b). The block centered finite difference method differs from an algorithmic point of view strongly from the face centered finite element method. However, for not too large grid bocks there is near mathematical equivalence between the two methods (Weiser and Wheeler 1988). More specifically, in the face centered method the fluxes within the grid blocks are linear interpolations between the fluxes through two opposite grid block faces and in the block centered finite difference method this is approximately the case too. Therefore, we base our analysis on the fluxes through faces $I - \frac{1}{2}$ and $I + \frac{1}{2}$ in the x-direction, the fluxes through faces $J - \frac{1}{2}, J + \frac{1}{2}$, in the y-direction, and the fluxes through faces $K - \frac{1}{2}, K + \frac{1}{2}$ in the z-direction. Considering the x-direction and omitting the subscript x (i.e., writing $q_x = q$) linear interpolation between the fluxes through two opposite the grid block faces results in

$$q(x) = ax + b$$
$$a = -\frac{q_{I-\frac{1}{2},J,K} - q_{I+\frac{1}{2},J,K}}{\Delta x_L} \tag{6.8}$$
$$b = \frac{q_{I-\frac{1}{2},J,K} + q_{I+\frac{1}{2},J,K}}{2}$$

where $x = 0$ denotes the grid block center. Integration results in

$$\int\limits_{-\frac{1}{2}\Delta x_I}^{+\frac{1}{2}\Delta x_I} q^2(x)dx = \frac{1}{12}a^2\Delta x_I^3 + b^2\Delta x_I \qquad (6.9)$$

We now choose the weighting function for flow in the x-direction equal to

$$w_{I,J,K}(x,y,z) = 1/\Delta V_{I,J,K} = 1/\Delta x_I \Delta y_J \Delta z_K \qquad (6.10)$$

From Eqs. 6.8, 6.9, and 6.10, we then find

$$\iiint\limits_{\Delta V_{I,J,K}} q^2(x)wdV = \overline{q^2}$$

$$= \frac{(q_{I-\frac{1}{2},J,K} + q_{I+\frac{1}{2},J,K})^2}{4} + \frac{(q_{I-\frac{1}{2},J,K} - q_{I+\frac{1}{2},J,K})^2}{12} \qquad (6.11)$$

If $q_{L-\frac{1}{2},M,N} \approx q_{L+\frac{1}{2},M,N}$, the second term at the right-hand side of Eq. 6.11 is sufficiently small to be neglected with respect to the first term in the right-hand side, which results in the approximation $(\overline{q^2})^{1/2} \approx |\bar{q}_x|$ with $\bar{q}_x = \left(q_{L-\frac{1}{2},M,N} + q_{L+\frac{1}{2},M,N}\right)/2$, as introduced in Chap. 3 and applied in Chap. 5. However, in grid blocks with sources or sinks the fluxes through two opposite faces may have opposite directions, $q_{L-\frac{1}{2},M,N} \approx -q_{L+\frac{1}{2},M,N}$, which results in an intolerably poor approximation when omitting the second term at the right-hand side of Eq. 6.11. For that case, we have to apply the complete Eq. 6.11, which can also be written as

$$\overline{q^2} = \frac{q_{I-\frac{1}{2},J,K}^2 + q_{I-\frac{1}{2},J,K}q_{I+\frac{1}{2},J,K} + q_{I+\frac{1}{2},J,K}^2}{3} \qquad (6.12)$$

For nonzero fluxes, Eq. 6.12 produces always positive values. For instance, if $q_{L-\frac{1}{2},M,N} = -q_{L+\frac{1}{2},M,N} = q$, we find $\overline{q_x^2} = q^2/3$. Similar expressions hold for flow in the y- and the z-directions, as well as for the components $q_x^2/\eta_{xz}, q_y^2/\eta_{xz}$ as they occur in Eq. 6.6.

If zone L contains only one grid block, i.e., if $L = n$, the grid block averaged terms of the form $|\overline{q_{n;i}}| = [(\overline{q_{n;i}})^2]^{1/2}$ in Eqs. 3.7 and 3.8 may be replaced with the terms $(\overline{q_{n;i}^2})^{1/2}$ resulting in

$$k_{n;i+1} = k_{n;i}\sqrt{\frac{\overline{(q_{n;i}^F)^2}}{\overline{(q_{;i}^H)^2}}} \qquad (6.13)$$

The update rules for the conductivities in the y- and z-directions can be derived in a similar way. Using this equation, the introduction of a small flux q^ε, as proposed in Sect. 3.4. for Eqs. 3.7 and 3.8, can be avoided.

6.4 Case Study Schietveld

In this section, we summarize the results of a case study presented in great detail by El-Rawy et al. (2017). Here we pay only attention to some theoretical aspects presented in Sects. 6.2 and 6.3.

A comprehensive hydrogeological description of the Schietveld basin has been presented by Batelaan et al. (2012), El-Rawy (2013), Nagels et al. (2015). Summarizing, the water table of the Schietveld basin has declined over several decades due to an enlarged number of drainage canals, leading to deterioration of protected heathlands. In order to set up a plan for nature restoration, a modeling study was required. The model area, which covers the Schietveld basin as well as the neighboring agricultural land, has a rectangular area of about 90 km^2, located on the Western side of the Campine Plateau in North–East Belgium. The elevation ranges between 50 and 182 m above sea level. This area is located within two major catchments that of the River Scheldt to the west and of the River Meuse to the east. The water divide runs approximately from north–east to south–west. The area includes several small streams and ditches that drain the site. Some depression bogs in the area are ecologically highly valued. The subsurface consists of Quaternary sediments and Tertiary fine sands forming a phreatic aquifer of 11–130 m thick. Below this, aquifer is a more poorly conducting aquifer, 100–220 m thick, composed of Tertiary medium to coarse sands resting on a very thick clay layer forming a natural no-flow bottom boundary. The two aquifers are connected, except in the northwest where they are separated by a clayey poorly conducting layer of 10–20 m thickness. Only general information concerning the properties of the hydrogeological layers, in particular their conductivity, is available from the above-mentioned regional groundwater modeling studies, as well as from pumping tests, but none of them are located inside the study area. In this model study, this information was considered as a source of "soft data." The boundary conditions are groundwater recharge at the water table and imposed heads at the four horizontal boundaries. The WetSpass water balance model (Batelaan and De Smedt 2007) was used to obtain a time-averaged (20-year period) spatially distributed recharge rate. The model requires spatially distributed data on elevation, land use, soil type, slope, and initial groundwater levels, as well as long-term averaged climate data obtained from the Kleine Brogel meteorological station, which is about 20 km north of the study area. The small streams and drains are simulated with the MODFLOW DRAIN package, imposing a drain level and conductance. Also, 92 pumping wells, mainly in the phreatic aquifer, are incorporated as internal boundary conditions with fixed abstraction rates ranging from 45 to 124,000 m^3/year.

By taking data from a period $\Delta t = 20$ year and time-averaging the equations, the conductivity estimation becomes considerably simpler. Doing so, the time-averaged specific storage term $s\Delta h/\Delta t$ in Eq. 2.4 and the time-averaged specific yield term $\theta\Delta h/\Delta t$ in Eq. 2.7 (the linearized approximation of Eq. 2.5) become negligibly small resulting in a steady-state model. Neglecting the model's nonlinearities (e.g., the appearance/disappearance of surface waters in wet/dry periods) the grid block conductivities determined by the DCM from the steady-state model are equal to the grid block conductivities that would be obtained by averaging of the time series of conductivities obtained by a sequence of DCM-observations (20 observations/year × 20 year = 480 observations). Since the average of the time series (Eq. 4.7) is practically equal to the Kalman-estimated conductivities (see the derivation of Eq. 4.13 from Eqs. 4.9 and 4.11), the uncertainty is relatively small with respect with the observation error (i.e., $\sigma_{n;M}/s_n^0 = 1/\sqrt{480} \approx 5\%$ if there are no model errors; see Eq. 4.14). However, the observation error s_n^o (Eq. 4.8) cannot be determined from an averaged steady-state model.

The steady-state MODFLOW-2005 model (Harbaugh et al. 2000; Harbaugh 2005) was built with support of the ModelMuse (Winston 2009) graphical user interface. The model has a horizontal resolution of 10 m × 10 m, or 785 rows (grid blocks in y-direction) and 1180 columns (grid blocks in x-direction), while three layers of variable thickness (z-direction) discretize the vertical direction. As a result, the model consists of 2,778,900 grid blocks. The initial horizontal layer conductivities were taken from the above-mentioned regional studies while the initial vertical hydraulic conductivities for all layers were based on an anisotropy ratio of 10.

In previous work (El-Rawy 2013; El-Rawy et al. 2015, 2016), the conductivities were estimated using the double constraint method in which each grid block in the model domain was considered as a zone. El-Rawy (2013) and El-Rawy et al. (2015) used the update rule given by Eq. 5.1, in which the grid block fluxes were obtained by averaging the face-based fluxes, like in Chap. 5, while El-Rawy et al. (2016) used the update rule given by Eq. 6.13. The results showed that only conductivities in the neighborhood of the head measurements (the measurement ranges) could be estimated, while in grid blocks at distance from the measurements (the terra incognita) DCM was ineffective. This is no surprise, because in an approach in which each grid block is a zone the grid block conductivities are not correlated to their neighbor grid block conductivities. However, using the available "soft hydrogeological" information about hydrogeological units, in which the conductivities are spatially correlated, we may introduce a zonation in which only a limited number of units is considered as a zone. In this way, the number of unknown conductivity values is drastically reduced. Inverse modeling based on a limited number of zones is relatively popular among practice-oriented hydrogeologists. Moreover, this approach can be handled by conventional gradient methods without leading to excessive computational requirements. In this Schietveld study, we have compared our DCM results with results obtained by the UCODE inversion model (Hill and Tiedeman 2007; Poeter et al. 2005).

In this particular Schietveld case, the zone-integrated double constraint method is applied considering four conductivity zones with independent horizontal and vertical anisotropic conductivities. This means that we have used the following two update rules (see Eqs. 6.4 and 6.6)

$$k_{H;L;i+1} = k_{H;L;i} \sqrt{\frac{\iiint_{D_L} [(q_{x;i}^F)^2 + (q_{y;i}^F)^2] w dV}{\iiint_{D_L} [(q_{x;i}^H)^2 + (q_{y;i}^H)^2] w dV}} \qquad (6.14)$$

$$k_{V;L;i+1} = k_{V;L;i} \sqrt{\frac{\iiint_{D_L} (q_{z;i}^F)^2 w dV}{\iiint_{D_L} (q_{z;i}^H)^2 w dV}} \qquad (6.15)$$

where in each grid block weighting function w is chosen equal to $w = 1/\Delta x \Delta y \Delta z$. In accordance with the theory presented in Sect. 6.3 of this chapter, this leads to Eq. 6.3 for the evaluation of the integrals in Eqs. 7.1 and 7.2.

Zone 1, the first layer of the model, represents the upper aquifer; zone 2a is the poorly conducting layer in the north–east; zone 2b is the remaining part in the second layer of the model; and zone 3, the third layer, represents the lower aquifer. Although zone 2b is expected to have similar properties as zone 3, it is considered as a separate zone to test the robustness of model calibration procedure, which is a rather tough test as there are no head observations in this zone. Hence, there are eight model parameters to be calibrated: $k_{H;1}, k_{V1}, k_{H;2a}, k_{V;2a}, k_{H;2b}, k_{V;2b}, k_{H;3}$, and $k_{V;3}$, for which general and impartial initial values have been chosen to start the DCM calibration procedure, i.e., 5 m/d for all conductivities, except for the poorly conducting zone 2a for which a value of 0.1 m/d was assumed. A run of the model with these initial conductivity values yields estimated heads that are in the order of about 5 m lower than the observations, which indicates that the initial horizontal conductivity values are too high and far from the optimum values.

The model was first calibrated in a classical way using UCODE (Hill and Tiedeman 2007; Poeter et al. 2005) and ModelMate (Banta 2011). However, after about 100 iterations and several hundreds of model runs, the procedure terminated without finding optimum parameter values. Obviously, the reason is that we are trying to estimate 8 parameters with only 37 observations. This is also reflected by the parameter sensitivities. A sensitivity analysis showed that $k_{H;3}$ is the most sensitive parameter, while $k_{H;1}$ and $k_{H;2b}$ are about 10 times less sensitive, while the remaining parameters are about 100 or more times less sensitive. Hence, with the present data it is impossible to estimate all parameters with UCODE unless additional constraints are imposed on the insensitive parameters. We opted to fix the vertical conductivities of zones 1, 2b, and 3 to one-tenth of the corresponding horizontal conductivities and to fix the horizontal and vertical conductivities of zone 2a to a small value of 0.001 m/d (i.e., virtually impervious). The remaining three parameters $k_{H;1}, k_{H;2b}$, and $k_{H;3}$ were calibrated with UCODE, and convergence (change in conductivity values less than 0.01) was obtained after 12 iterations and 84 model runs.

Using DCM to determine these three parameters we applied Eq. 6.6 with $\eta_{xz} = \eta_{yz} = 10$ for zones 1, 2b, and 3. Convergence (change in conductivity values less than 0.01) was obtained after a first DCM stroke followed by 11 iterations (i.e., after 24 model runs). However, it turned out that application of DCM using Eqs. 6.14 and 6.15 to determine the above-defined eight conductivities does not give problems and results in realistic conductivities and anisotropy ratios, as can be observed from the comparison presented in Table 6.1.

In Table 6.1, HC means hydraulic conductivity, INIT means initial value in m/d, UCODE and DCM denote the two inversion techniques, (a) means fixed anisotropy ratio $k_H/k_V = 10$, (f) means fixed conductivity value. The numbers between square brackets, [...], denote the 95% confidence intervals in m/d derived under the assumption that the estimated conductivities have lognormal probability distributions.

The thus-obtained conductivity values and confidence intervals determined by UCODE are reasonable, except maybe for $k_{H;2b}$ as this value is rather high and the confidence interval rather wide. Moreover, the value of $k_{H;2b}$ deviates considerably from $k_{H;3}$ and the confidence intervals do not overlap, although these conductivities relate to the same aquifer. However, the parameters are strongly linked because the correlation coefficient determined by UCODE is −0.89. Hence, if one of these horizontal conductivities increases, the other one would decrease so that the overall conductivity of the lower aquifer is less affected. The conductivities determined by UCODE yield a mean absolute error (MAE) and a root mean squared error (RMSE) between observed and simulated heads equal to 0.21 and 0.48 m, respectively. When UCODE is rerun with other starting values almost the same optimum values and confidence intervals are obtained, hence we can conclude that the global optimum has been attained.

For the inversion by DCM, we determined the eight horizontal and vertical conductivities of the four zones using Eqs. 6.14 and 6.15. To terminate the iterations, we impose that the conductivity updates should be larger than 1%; that is, we accept that the method has converged when for each zone $\Delta k/k \leq 0.01$. Unlike UCODE the DCM runs without any problem with little computational effort. During the iterations, the conductivities vary rather smoothly from their initial value and tend asymptotically to their final value. The largest variation occurs for $k_{H;1}$, which is also the parameter with the slowest convergence rate. During the iterations, we noticed an almost exponential decrease of the relative conductivity

Table 6.1 Comparison between UCODE and DCM

HC	$k_{H;1}$	k_{V1}	$k_{H;2a}$	$k_{V;2a}$	$k_{H;2b}$	$k_{V;2b}$	$k_{H;3}$	$k_{V;3}$
INIT	5	5	0.01	0.01	5	5	5	5
UCODE	0.35 [0.10–1.21]	0.035 (a)	0.001 (f)	0.001 (f)	19.8 [7.6–51.4]	1.98 (a)	0.98 [0.42–2.30]	0.098 (a)
DCM	0.28	0.54	0.025	0.033	2.15	0.75	2.4	1.78

changes. The relative change in the conductivity of most zones becomes smaller than 0.01 after relatively few (about 20–30) iterative DCM strokes (i.e., about 40–60 model runs), except for $k_{H;1}$, which needs one first DCM stroke followed by 57 iterations (116 model runs) to converge. This slower convergence is caused by the fact that most head observations are situated in the first model layer (zone 1, the upper aquifer). This is less important for the vertical conductivity because the groundwater flow in the study area is largely horizontal. The fastest convergence rate was noticed for $k_{H;3}$, likely because layer 3 is the lower, largest aquifer where most of the groundwater flow occurs. All estimations are within reasonable orders of magnitude. Striking is the fact that the vertical conductivities have the same order of magnitude for all zones, which means that the anisotropy ratio is not that high as previously anticipated. Nevertheless, it must be stressed that the groundwater flow is predominantly horizontal so that the vertical conductivities are less important and have little impact on the numerical model, which likely gives rise to the uncertainty of their estimates. The horizontal conductivity values for zone 1 and zone 3 estimated by DCM comply with what was obtained by UCODE, but this is not the case for zone 2b. With DCM the conductivities of zones 2b and zone 3 are found to be very similar, which agrees with what is known from the geological buildup of these layers. However, with UCODE the conductivity of zone 2b becomes rather high and somewhat unrealistic.

Regarding the variation of the RMSE of the differences between observed and simulated heads obtained with the flux model during the DCM iterations, we observed a very smooth decrease and asymptotic convergence. The final MAE and RMSE values are −0.68 and 0.78 m, respectively, which is somewhat lower than what was obtained by UCODE.

Regarding the simulated heads versus the observed heads, we notice that the simulated heads obtained from UCODE are in better agreement with the observations from the 37 piezometers, although the results obtained with DCM are almost of the same quality. The difference may be explained by the fact that UCODE is a calibration method; that is, it is a method based on minimization of the differences between observed heads and heads obtained from the flux model. In contrast to UCODE, DCM is an imaging method minimizing the errors in the Darcy equation, in which the fluxes are obtained by the flux model, like in UCODE, while the observed heads are matched by an auxiliary model: the head model (see Chap. 3, Sect. 3.9.1). As has been explained in Chap. 3, Sect. 3.9, models based on MODFLOW under estimate the conductivities. As a consequence, the flux model, although yielding the correct fluxes, over-estimates the head gradients. These considerations explain the difference between UCODE (better match with the measured heads) and DCM (better match with the real field conductivities). Moreover, the heads imposed on the head model (measured in piezometers) and the fluxes imposed on the flux model (recharge, well rates) may introduce systematic errors, even if the random errors have decreased by the time-averaging over 20 years. These errors may cause that DCM minimization of the Darcy errors does not make the Darcy residual (Eqs. 6.1 and 6.2) equal to zero.

In conclusion, the results obtained by DCM are good and, from a practical point of view to be preferred, because DCM enables to obtain estimates for all horizontal and vertical conductivities of the zones without any additional assumptions or restrictions and with no more effort or complication than a classical gradient-based calibration method.

References

Banta ER (2011) ModelMate—a graphical user interface for model analysis. US Geological Survey Techniques and Methods, Book, p 6

Batelaan O, De Smedt F (2007) GIS-based recharge estimation by coupling surface-subsurface water balances. J Hydrol 337:337–355

Batelaan O, El-Rawy M, Schneidewind U, De Becker P, Herr C (2012) Doorrekenen van maatregelen voor herstel van vochtige heidevegetaties op het Schietveld van Houthalen-Helchteren via grondwatermodellering (Scenario analysis and groundwater modeling for the rehabilitation of wet heathlands at Houthalen-Helchteren military domain). Final Rep., Brussels, Belgium

Calderón AP (1980) On an inverse boundary value problem. Seminar on Numerical Analysis and its Applications to Continuum Physics (Rio de Janeiro 1980), Soc Brasil Mat, 65–73, Rio de Janeiro. Also see the reprint: Calderón AP (2006) On an inverse boundary value problem. Comput Appl Math 25(2–3):133–138

Chavent G, Jaffré J (1986) Mathematical models and finite elements for reservoir simulation. Elsevier, North-Holland, Amsterdam

De Smedt F, Zijl W, El-Rawy M (2017) A double constraint method for analysis of a pumping test. Submitted

El Rawy M, De Smedt F, Zijl W (2017) Zone integrated double constraint methodology for calibration of hydraulic conductivities in grid cell clusters of groundwater flow models. Submitted

El-Rawy M (2013) Calibration of hydraulic conductivities in groundwater flow models using the double constraint method and the kalman filter. Ph.D. thesis Vrije Universiteit Brussel, Brussels, Belgium

El-Rawy M, Batelaan O, Zijl W (2015) Simple hydraulic conductivity estimation by the Kalman filtered double constraint method. Groundwater 53(3):401–413. https://doi.org/10.1111/gwat.12217

El-Rawy M, De Smedt F, Batelaan O, Schneidewind U, Huysmans M, Zijl W (2016) Hydrodynamics of porous formations: simple indices for calibration and identification of spatio-temporal scales. Mar Petrol Geol 78:690–700. https://doi.org/10.1016/j.marpetgeo.2016.08.018

Harbaugh AW (2005) MODFLOW-2005, The U.S. geological survey modular ground-water model, the ground-water flow process. Techniques and Methods-6-A16, US Geological Survey, Reston, VA

Harbaugh AW, Banta ER, Hill MC, McDonald MG (2000) MODFLOW-2000, The US Geological Survey modular ground-water model, User guide to modularization concepts and the ground-water flow process. US Geological Survey Open-File Report 00-92, 121, Reston, VA

Hill MC, Tiedeman CR (2007) Effective groundwater model calibration: with analysis of data, sensitivities, predictions, and uncertainty. Wiley, New York

Kaasschieter EF (1990) Preconditioned conjugate gradients and mixed-hybrid finite elements for the solution of potential flow problems. Ph.D. thesis Delft University of Technology, Delft, Netherlands

Kaasschieter EF, Huijben AJM (1992) Mixed-hybrid finite elements and streamline computation for the potential flow problem. Numer Methods Partial Differ Equat 8:221–266

Nagels K, Schneidewind U, El-Rawy M, Batelaan O, De Becker P (2015) Hydrology and ecology: how Natura 2000 and Military use can match. Ecol Questions 2:79–85

Poeter EP, Hill MC, Banta ER, Mehl S, Christensen S (2005) UCODE-2005 and six other computer codes for universal sensitivity analysis, calibration, and uncertainty evaluation. US Geological Survey Techniques and Methods 6–A11, Reston, VA

Weiser A, Wheeler MF (1988) On convergence of block-centered finite differences for elliptic problems. SIAM J Numer Anal 25:351–375

Winston RB (2009) ModelMuse-A graphical user interface for MODFLOW-2005 and PHAST. Techniques and Methods-6-A29, US Geological Survey, Reston, VA. http://pubs.usgs.gov/tm/tm6A29

Zijl W (2005a) Face-centered and volume-centered discrete analogs of the exterior differential equations governing porous medium flow I: theory. Transp Porous Med 60:109–122

Zijl W (2005b) Face-centered and volume-centered discrete analogs of the exterior differential equations governing porous medium flow II: examples. Transp Porous Med 60:123–133

Chapter 7
Summary and Conclusions

We have demonstrated a double constraint methodology for the determination of hydraulic conductivities in grid volumes of groundwater flow models. The methodology is straightforward as it allows to update hydraulic conductivities by simply comparing the results of a model run in which known fluxes are imposed as boundary conditions with the results of a model run in which known heads are imposed as boundary conditions. In this approach, flux and head conditions in wells —including zero-flux observation wells—are considered as boundary conditions too. In the zone-integrated double constraint approach, the model domain is partitioned into zones with presumed constant hydraulic conductivity. In this sense, the original double constraint method is a zone-integrated double constraint method in which each zone is a grid volume.

The feasibility and practical use of the (zone-integrated) double constraint methodology has been illustrated by two practical case studies, the Kleine Nete case in which each grid block in the MODFLOW model is a zone, and the Schietveld case in which only four zones are considered. In the latter case—the four zones case —the horizontal and vertical conductivities determined by double constraint methodology compared favorably with the results obtained by UCODE, which is based on a classical gradient method that minimizes the differences between calculated and measured heads. Since the classical approach needed additional assumptions and simplifications to reach a meaningful result, we may conclude that the double constraint methodology is more robust. Moreover, in contrast to classical gradient-based methods, the number of zones in the double constraint methodology can simply be extended to any number of zones less than or equal to the number of grid blocks, without excessive increase in computation time and computer memory requirements, as has been demonstrated by the Kleine Nete case study.

Because the double constraint methodology is not only based on the model of interest for the study (a flux model), but also on an auxiliary model (the head model), it is not a calibration method in the strict sense; in fact, it is an imaging method. Its results may differ a bit from an exactly calibrated model obtained by an error-free calibration method. However, in practice observations are not error-free,

© The Author(s) 2018
W. Zijl et al., *The Double Constraint Inversion Methodology*,
SpringerBriefs in Applied Sciences and Technology,
https://doi.org/10.1007/978-3-319-71342-7_7

as has been demonstrated and quantified by the Kleine Nete case study. There will always be random and systematic differences and mismatches between observations and model predictions so that an exact model calibration can never be obtained.

In view of the fact that determination of the parameters in a groundwater model is a nonlinear problem and conventional gradient-based inversion by optimization (minimization) techniques are computationally cumbersome and time-consuming, and accepting that parameter determination needs not to be more accurate than justified by the accuracy of the head or flux observations, we may conclude that the double constraint methodology produces sufficiently reliable results at relatively low computational cost. Hence, the double constraint methodology proves to be a promising approach for the estimation of parameter values of groundwater flow models (especially hydraulic conductivities).

Comparison with a conventional gradient method, presented in Sect. 6.4 by the Schietveld case study, suggests to devote here some thoughts about the similarities and differences between the conventional methods and the double constraint methodology. Although the comparisons presented in Chap. 6 show that the results do not differ too much, i.e., are within the range of measurement errors, experiences obtained by only two comparisons cannot be decisive to make a choice between the different methods. Therefore, we will present here some arguments to weigh the merits of the different approaches. For problems in which "big data" play a role the ensemble Kalman filter (EnKF), the steady Kalman filter (SKF) and the double constraint methodology (DCM) might be the better choice, while for problems handling only relatively few data conventional gradient-based optimization techniques might be preferable. Often the choice has to be based on the more practical question whether standard software is easily available. Although the double constraint methodology is relatively simple to implement (just divide the results of two MODFLOW models), the criterion of ready-to-use software often forces the hydrogeologist to routinely apply UCODE or PEST (conventional gradient methods). In the petroleum reservoir community, the present-day ready-to-use choice may also be the ensemble Kalman filter. For a detailed explanation of the ensemble Kalman filter (EnKF) inversion technique, see Sect. 4.3.2 of Chap. 4.

There are, however, also more theoretical—or, if you like, more philosophical—ways to discuss the similarities and the differences. The first remark is that, in our opinion, the popular classification in direct and indirect inversion methods is not very helpful. In fact, we have given arguments why the distinction between calibration methods and imaging methods is more relevant. More specifically, the double constraint methodology is an imaging method, while the conventional inversion techniques are calibration methods (for a detailed explanation of the difference between imaging and calibration see Sect. 3.9 of Chap. 3). This means that the results obtained from the double constraint methodology will always be different from the results obtained by conventional methods. For comprehensive numerical models—models based on a fine-scale grid—these differences can be made sufficiently small by grid refinement of the model, but for simple models—models based on a coarse grid—these differences may become significant. Objective criteria upon which a choice between calibration and imaging can be based are not (yet) available; the best we can say is that the two approaches are complementary rather than competing; also see Sect. 3.9.2 of Chap. 3.

Although the double constraint methodology can be used as a "stand-alone method" to determine parameters, especially conductivities, the method can also be applied in combination with other inversion techniques. In Sect. 4.3.3 of Chap. 4, suggestions have been presented how the ensemble Kalman filter (EnKF) can be assisted by the double constraint methodology (DCM) to overcome over-smoothing by the EnKF. If a conventional gradient methodology is preferred, the double constraint methodology could be used to precondition the conventional method. If the initial guess of the conductivities is far removed from the true conductivities, the first stroke of the double constraint methodology—without subsequent iterations—gives already a good estimate of the orders of magnitude of the conductivities. Thanks to this property the double constraint methodology can be used as a simple tool to find conductivities that are acceptable for (are within the domain of convergence of) the conventional methodology. In addition, using the double constraint methodology gives insight in the physics of flow contained in its governing equations. For instance, from the understanding of the double constraint methodology, it is immediately clear that conductivity estimation is meaningful only if there are more conditions imposed on the boundaries and in the (observation) wells than required for a forward problem. In other words, if conditions that have to be imposed for a forward problem are unknown, they cannot be found by inverse modeling.

In Chap. 4, Sect. 4.2, it has been shown that thanks to Kalman filter post processing, repeating the DCM estimations for varying hydrological conditions results in conductivities with lower uncertainty than the measurement errors. This uncertainty reduction has been demonstrated in Chap. 5 by the Kleine Nete case study, while a comprehensive explanation of the Kalman filter is presented in Chap. 4, Sect. 4.3.

Finally some words about the justification of our approach to mathematics. The Calderón problem was one of the sources of our inspiration. Roughly speaking, this problem includes all aspects of Calderón's conjecture stating that, under some smoothness conductions, the conductivity distribution in the model domain can be uniquely determined by imposing both flux and head on the closed boundary of the model domain. However, the literature on this conjecture is generally using overly rigorous mathematical tools based on infinite function spaces. We have avoided this mathematical type of analysis, not only because it is hard to understand for practice-oriented geoscientists and engineers, but also because this type of analysis is too abstract to suggest practical constructions of solutions. Therefore, rather than basing our approach on the concept of infinity from the beginning, we have presented our analysis in a more insightful way by basing it on discretized, or quantized problems, like in numerical models with a large number of small grid cells (edges, faces, volumes). Although we have used the powerful mathematical tools of classical analysis (differentiation, integration, differential equations, etc.), the concept "infinity" has been used only in the meaning of an extremely large number of extremely small grid cells, without the need to specify how much "extremely large" and how small "extremely small" actually is. Anyhow, the thus-conceived "infinitesimally small" volume is larger than a representative elementary volume of the porous medium.

Printed in the United States
By Bookmasters